weight

KNOWLEDGE APPLICATION

KNOWLEDGE
APPLICATION

*the knowledge system
in society*

BURKART HOLZNER
University of Pittsburgh

JOHN H. MARX
University of Pittsburgh

Allyn and Bacon, Inc.
Boston   London   Sydney   Toronto

To the memory of
Paul F. Lazarsfeld and Edward A. Suchman,
friends and colleagues

Library of Congress Cataloging in Publication Data

Holzner, Burkart, 1931–
    Knowledge Application.

    Bibliography: p.
    1. Knowledge, Sociology of. 2. Social change.
3. Science and civilization. 4. Technology and
civilization. I. Marx, John H., joint author.
II. Title.
BD175.H33   301.2'1   78-234-77
ISBN 0-205-06516-3

Printed in the United States of America

Contents

As the varying constitution of the speculative class necessarily represents the corresponding situation of the human mind in general, the nascent positivism of the last three centuries has given to the mathematicians more and more of that authority which, till the end of the medieval period, had belonged to moral and social researches. This provisional anomaly will now come to an end; for, when sociological theory has once reached the positive state, there is nothing except the opposition of the ignorant and the interested, to prevent the human view from resuming its natural place at the head of all human speculation.

Auguste Comte

Specialists without spirit, sensualists without heart; this nullity imagines that it has attained a level of civilization never before achieved.

Max Weber

Speaking for myself, I too believe that humanity will win in the long run; I am only afraid that at the same time the world will have turned into one huge hospital where everyone is everyone else's humane nurse.

Goethe

Acknowledgments

This book grew out of a course on the sociology of knowledge application that we taught both jointly and separately as one component of the University of Pittsburgh's applied sociology program. We owe much to the vigorous discussions with our students, undergraduate and graduate—some of them accomplished professionals wishing to learn about the intricacies of the knowledge system. Our colleagues at the University have stimulated ideas and we have benefited especially from three conferences: the 1973 conference on knowledge utilization organized by the late Paul Lazarsfeld as a cooperative Columbia-Pittsburgh venture; the 1977 conference on social research organizations convened by Jiri Nehnevajsa and Robert Avery in Pittsburgh; and in 1975 Roland Robertson and Burkart Holzner held a small working conference in Pittsburgh on identity and authority that helped us focus on the relations between these concepts and knowledge. Rainer Baum, Evelyn Fisher, Bernhardt Lieberman, and Leslie Salmon-Cox have helped through criticism and suggestions.

We are especially indebted to the secretarial staff of our department; we hardly dare hope that the clarity of our ideas matches the excellence of their work. Janet Claherty and Michele Murphy have been patient, hardworking, and precise. Dorothy Lederman has managed both our lives and our work—and her friendship has given us encouragement.

Our families have put up with a lot—love and thanks to Anne Holzner and Nancy Marx.

Burkart Holzner
John H. Marx

Preface

The central theme of this book concerns the knowledge system. We address this theme, focusing on knowledge application, from the perspective of the social construction of reality. In adopting this point of view we do not advocate a simplistic relativism, nor do we assert that reality is a mere social construction—quite the contrary. We believe what we have to say is of both basic theoretical and concrete practical significance. Concern with the role of knowledge and expertise in contemporary society is widespread. Yet there are few works that offer a systematic theoretical perspective for its understanding and interpretation. Most of the literature dealing with the problem of knowledge application is written from the limited, highly specialized point of view of administrative or management concerns. While not denying the utility of such work, we are convinced that a broader framework is needed, grounded in a basic sociological theoretical perspective. We hope that our frame of reference will prove useful in relating heretofore segregated, more specialized empirical or practical perspectives and problems to each other.

The study of the role of knowledge in contemporary society has been conducted from many quite specific and often unrelated points of view. Because this is so and because we wish to show the fruitfulness of a general perspective, we conceive of the abstractness of certain portions of our book as a strength, rather than a weakness. This abstractness should make it possible to apply the ideas and point of view of this book in seemingly disparate substantive and empirical arenas whose interrelation it may make apparent. Of course, we address an audience of those interested in the sociology of knowledge and sociological theory, but we also hope to find interest among intellectuals generally and among practitioners in the personal service professions, applied sociologists, policy analysts, social and urban planners, and businessmen in marketing and management fields.

The very nature of our subject matter, the contemporary knowledge system, also requires that we encompass an unusually diverse range of topics and issues in our text. Thus, we touch on work in management, marketing, and diffusion of innovations; technology and society; professional organization, socialization, and communities; symbolism, common sense, and ideology; as well as the history and sociology of science and issues in the classical sociology of knowledge. This scope is necessary because of the topic, problems of knowledge production, distribution, utilization, transfer, and storage.

These considerations have influenced our decisions about certain matters of presentation. We have attempted to refrain from unnecessary jargon—even though others may dispute whether the technical terms we find essential are not jargonistic, after all. Yet we have tried to present our ideas in readable form and thus have, for example, avoided presenting any of the numerous statistical tables that support our arguments in chapters 1, 6, 7, and 8 concerning *post-modern society* and the production, organization-distribution, application, and storage of knowledge in contemporary society. We made this decision to avoid interrupting the flow of the text and in order to focus on the development of the argument itself. For the same reason, we have avoided any substantive footnotes and scrupulously attempted to delete tangential digressions, however tantalizing to us, from the text. There are obviously both advantages and drawbacks to this approach. In order to provide some of the bases for our arguments, we have included numerous references; in addition, we offer brief selected bibliographies at the end of each major section. The purpose of these section bibliographies is triple: we hope the selection of books will be of use to interested readers for further study; they also include the central references that we have drawn on in each section so that the skeptical reader may check the validity of our interpretation; and we have tried to choose editions of original texts that are readily accessible in major urban libraries. For this last reason, we have also included some readers and anthologies.

Finally, we should point out that this book, like many recent publications on social science themes, eludes classification into traditional categories like teaching textbook, specialized scholarly monograph, or scientific study. In order to treat its subject adequately, we have had to raise a number of technical and complex issues and we have treated them seriously. To the extent that we have been able to render these issues available to thoughtful readers both in the sociology of knowledge and outside of its narrow confines, we will have succeeded in at least part of our intentions in writing this book.

Introduction

This book is about knowledge in contemporary society. It represents an attempt to describe and analyze certain kinds of realities that are taken as "known" by members of certain specialized subcultures. In one sense, it is part of a slowly growing body of work in the sociology of knowledge that is concerned with the social construction of cultural meanings, maps, and models that define and determine what we take for granted as reality. However, this book is about one particular kind of knowledge that is especially important in contemporary society; namely, specialized systematic, technical knowledge—such as that associated with science, technology, or professional practice. Moreover, our concern lies with the manner in which the production, organization, distribution, application, and utilization of specialized, technical knowledge has transformed social life by creating a post-modern, knowledge-based society. Thus, the main focus of this treatise is on the application of knowledge in post-modern culture.

Sociologists seem to be obsessed with first labeling and then locating their perspective in relation to the labels others have coined before them, and we lack both the imagination and the courage to eschew this practice. Therefore, we conceive of this treatise as representing a departure from the traditional sociology of knowledge that emphasized the social determinants, conditions, and distortions of knowledge. This Marxian-Mannheimian tradition in the sociology of knowledge typically treated ideas, beliefs, perspectives—in short, knowledge—as the dependent variable, to be explained by various structural and positional independent variables associated with specified interests. The perspective offered here, on the other hand, typically examines knowledge as an independent variable and focuses on its consequences for either the encompassing social system or other aspects of the cultural system. Furthermore, our concern with the social processes and institutional structures, through which cultural models of and for what people experience as realities are constructed, forces us to grant a much larger role to the interplay between cultural symbol systems and individual cognitive processes than is typically the case in the sociology of knowledge. For all of these reasons, we conceive of the present work as located within the sociology of culture or, to indulge in really inelegant jargon, the cultural sociology of specialized, technical knowledge in post-modern society.

Our frequent use of the heavy-handed term *post-modern* also warrants some explanation since other sociologists, both serious and superficial, have employed analogous concepts—albeit in significantly different fashion. Specifically, a number of recent analyses have attempted to characterize contemporary American society as post-industrial, focusing on dramatic changes in the economic and technological bases of industry and the occupational structure. While not disputing the magnitude or significance of these changes in the structure of American society, our analysis suggests that the more significant developments have occurred in the cultural system, the realm of socially constructed meanings, models, and bodies of knowledge. For this reason, we use "post-modern" to refer to cultural symbols, models, and knowledge of all kinds that distinguish contemporary American society from earlier periods.

The central analytical construct guiding our discussion derives from the idea of the social system of knowledge or, more specifically, the post-modern knowledge system. What we have in mind is quite straightforward. The complexities of social reality can be analyzed from a variety of perspectives and purposes. For example, some analysts are interested in the production and distribution of power, authority, and influence; others in the channels and media that emerge from the socially structured flow of communications; and still others in the ideological and informational content of these mass communications. The idea of the power structure or the communications network or the mass media, as well as more familiar terms like the economic system, represents analytical constructs that guide such modes of dissecting social reality. Each of these modes of abstracting from reality is in some fashion interdependent with the others—one cannot imagine an intelligent analysis of societal power structures that ignores economic realities or the mass media of communication. Yet they also are analytically distinct and provide different conceptual foci of attention. In an analogous sense, we use the term "knowledge system" to refer to that particular aspect of the structure and functioning of post-modern societies that emerges when one focuses on the socially structured distribution of knowledge related processes.

This conception is by no means original: both Fritz Machlup and George Gurvitch have employed it, albeit from very different perspectives. Machlup (1962) approached knowledge from an economic standpoint, describing the "production and distribution" of knowledge in the United States in minute detail. His analysis treated knowledge as cognitive resources differentially produced and distributed throughout the social structure and, hence, differentially accessible to occupants of various status positions. Machlup's study provided one point of departure for the present work in emphasizing the importance of the way in which knowledge and the cultural processes related to knowledge are structurally distributed and organized. Gurvitch focused on social structures such as villages, cities, and societies as frameworks for patterning knowledge as well as repositories for storing knowledge. His macrosociological perspective drew attention to what we call the knowledge system by specifically emphasizing its embeddedness in the colorful contexts of historical communities, institutions, classes, and solidarities. Thus, Gurvitch's work provided another cornerstone for the present analysis.

We use the conception of a post-modern knowledge production and use system as a sensitizing and organizing concept. We suggest that in post-modern societies, knowledge systems emerge as empirically differentiated, in part deliberately planned, sets of interlocking institutions and roles. To facilitate analysis of this phenomenon, we distinguish between analytically different, but empirically interrelated, aspects of the knowledge system. First, there is the creation or discovery of historically or at least situationally new knowledge, which we term *knowledge production*. Next, there is the *organization of knowledge* into coherent bodies or cultural packages that have a certain degree of cohesiveness and consistency. The cultural processes that are involved in this aspect of the knowledge system are closely related to those that *disseminate knowledge and differentially distribute* it throughout the social structure. These processes, in turn, are linked to institutional patterns and structures for *knowledge storage and accessing.* The processes and structures involved in the *organization, dissemination-distribution,* and *storage-retrieval* of specialized, technical knowledge comprise a cultural domain that is analytically separable from the domain of knowledge production. Finally, there is the domain of *knowledge application* that involves putting specialized technical knowledge to use in solving concrete, practice problems. The borders of this domain shade off into the area of *implementation,* which involves the absorption of knowledge into everyday routines of explicitly knowledge-based plans for action. These distinctions deal with interdependent and frequently fused aspects of the empirical knowledge system. It would be absurd to argue that knowledge production is independent of the other functions and processes. Indeed, we subsequently will suggest that knowledge application is increasingly becoming intertwined with knowledge production processes, with potentially alarming consequences. Nevertheless, we consider it useful to distinguish analytically these aspects and processes, initially treating their structural location and cultural dynamics separately, in order to provide a comprehensive sketch of the knowledge system in post-modern societies.

We have divided the book into three parts. The first part presents some background considerations pertaining to the knowledge system and the sociological analysis of post-modern culture. The first of the three chapters that comprise Part I describes the new kind of sociocultural system that is emerging in the United States and other highly technically advanced societies. This chapter introduces the concept of post-modern culture and indicates its relation to other sociological terms that have been used to describe the emergence of the new kind of sociocultural order that distinguishes contemporary society from previous social systems. The second chapter focuses on a central issue of this emerging sociocultural system: the development of a coherent conception for the application of systematic, organized, specialized knowledge. This chapter explores the core aspects of the process of knowledge application in some detail and presents some historical models that have been proposed as frameworks for implementing, utilizing, and applying knowledge in the past. The third chapter in Part I presents an historical overview of the sociology of knowledge. The purpose of this overview is to demonstrate the need for different perspectives than those associated with the mainstream of traditional sociology of knowledge approaches in order to deal with the new kinds of problems and phenomena that are associated with the emergence of post-modern culture and knowledge systems. Thus, Part I indicates the broader sociocultural developments and societal changes that make the knowledge system and knowledge application critical foci for sociological analysis. It further suggests the need for new conceptual perspectives and approaches appropriate to the interpretation of knowledge-related developments in post-modern cultures.

The two chapters that comprise Part II are addressed to the task of developing the conceptual framework and tools with which we propose to analyze post-modern knowledge systems and processes of knowledge application. The first of these chapters presents the main conceptual tools of inquiry for our formulation. Although they represent the conceptual foundation underlying our subsequent discussion and interpretation of contemporary developments in post-modern society, we believe that they comprise critical theoretical elements for any systematic sociological analysis of knowledge, knowledge application, cultural systems, sciences and professions, and other general topics that involve specialized bodies of knowledge. This chapter also describes our perspective on the social construction of reality and its application to contemporary, knowledge-based social patterns in some detail. The second chapter in Part II, chapter 5, takes up the critical question of the relation between social structure and differentially distributed bodies of knowledge. The analysis of this relation provides the conceptual basis for exploring the analytically differentiated aspects or components of the post-modern knowledge system in Part III of the book.

In the third part of the book, we take up the analysis of the three aspects of the knowledge system: production; organization, distribution, and storage; and application. Chapter 6 discusses knowledge production, the creation or discovery of historically or situationally new knowledge. It argues that the production of knowledge has increasingly become closely associated with specific kinds of social structures and contexts. The chapter examines the historical development of these knowledge-producing structures and communities down through the present time. The central emphasis is on culturally defined models and normatively sanctioned organizational arrangements that facilitate knowledge production and growth.

Chapter 7 deals with what can be thought of as the linkages in the knowledge system; namely, the processes of knowledge organization, differential distribution, and storage-retrieval. In this chapter, we depart most radically from other writers who have discussed the role and nature of knowledge in contemporary society, since none of them devote systematic attention to these linkages and channels that we believe are critical elements in the systemlike character of post-modern knowledge processes and institutions. This chapter naturally strains against its boundaries. The web of such cultural and structural knowledge linkages, as well as the subjects of the distribution and storage of knowledge in post-modern society, is of such scope and magnitude that we can only provide hints and direction for further analyses, rather than do justice to the complexities of the subjects.

Chapter 8 on knowledge application explores the relations between the knowledge system and the manifold worlds of practical activity. This chapter is of especially strategic significance to our discussion of the contemporary knowledge system because we argue that post-modern culture selectively emphasizes and rewards applied knowledge and the activities that comprise knowledge application at the expense of other aspects of the knowledge system. Our treatment of the process of knowledge application departs from other approaches to the subject, which have typically reflected managerial or policy-planning needs and assumptions. Our treatment is explicitly rooted in the reality constructionist perspective of the sociology of knowledge and culture. This leads us to emphasize such topics as the linkages between social measurement, social movements, and knowledge application, for example, which other discussions of the application of knowledge to concrete (usually administrative or managerial) problems typically ignore. With this chapter, our analytical scheme comes full circle as it deals with the incorporation of specialized, technical knowledge into the ongoing life of the society. This chapter thus concludes Part III.

Part IV of the book consists of chapter 9, "Speculations on Sociocultural Change." This chapter indicates some broader connections between the themes discussed throughout the book and major foci of social and cultural change. The discussion brings us back to some of the general considerations and issues raised in Part I of the book. Specifically, we attempt to locate our analytically specialized interests in knowledge application in the context of some broader concerns about the shape, direction, and meaning of emerging social and cultural forms. Thus, we consider issues such as the impact of the contemporary knowledge system on directions and processes of sociocultural change and vice versa. These speculations lead us beyond the reassuring boundaries of what is or can be reliably known. This concluding chapter therefore is informed with a modest appreciation of the wide range of the possible, sharpened by a systematic skepticism about sociological forecasting and prophetic visions of the future.

KNOWLEDGE
APPLICATION

PART
I

Post-Modern Society and the Sociology of Knowledge

The definition of the situation [in modern society] is equivalent to the determination of the vague. In the Russian *mir* and the American rural community of fifty years ago nothing was left vague, all was defined. But in the general world movement to which I have referred, connected with free communication in space and free communication of thought, not only particular situations but the most general situations have become vague.

(*W. I. Thomas*, The Unadjusted Girl, *1923, pp. 81f.*)

The theme of this book is the social system of knowledge. It offers an analytical framework for an exploration of the social structural arrangements through which knowledge is produced or discovered, differentially distributed and disseminated, stored and retrieved, and applied and implemented. In particular, it attempts to demonstrate the crucial nature of the social embeddedness of the knowledge system as well as the significance of the knowledge system for historical change in culture and society. This focus is important not only to students of the sociology of knowledge, but also for anyone seeking an understanding of contemporary society that goes beyond superficial observations.

This first part of the book presents some critical back-ground considerations. Chapter 1 describes the new socio-cultural system that is emerging in the United States and other highly technically advanced societies. The chapter suggests that the emergence of post-industrial social structure based on advanced technology is accompanied by the growth of post-modern cultural systems based on the systematic application of specialized, technical knowledge to a wide range of human affairs. The discussion suggests that the emergence of post-modern knowledge systems has already begun to have a profound impact on the cultural domain of common sense.

The second chapter focuses on the process of knowl-edge application, the systematic application of specialized, technical knowledge to concrete problems. Several historically prominent models of and for applying theoretical or technical knowledge to practical affairs are examined in order to display the diverse approaches that continue to dominate current dis-cussions about the subject.

The third chapter in Part I presents an historical over-view of the sociology of knowledge. The purpose of this over-view is to indicate the need for a new perspective in the socio-logy of knowledge in order to deal more adequately with the kinds of knowledge problems and knowledge systems that are emerging in postmodern cultures. The chapter and first part of the book conclude with suggestions concerning new theo-retical perspectives for a sociology of knowledge and culture.

1

The New Society and
Its Common Sense

A NEW SOCIETY

The social order and culture of the United States have changed quite dramatically from the pattern of the industrial society of the era of the great World Wars. Between 1940 and 1970, the population of 132 million people increased to 205 million. While the rural population declined slightly from 57 million to 54 million, the urban population of 74 million in 1940 nearly doubled to reach a total of almost 150 million by 1970. In this same thirty-year period, the male labor force increased from 42 million to 54 million, and the female labor force expanded from 14 million to almost 32 million. Large as these numerical changes are, they only hint at the changes in the social structure during this period. Major change is evidenced in every institution from the family to the economy to the government. In 1940, there were 3,879,000 professional and technical workers in the country; in 1970, this category had grown to 11,561,000. The service occupations became increasingly important. Most significantly, there has been an enormous growth in the knowledge-related occupations and professions.

The keeping of statistics about a social process reflects an awareness of its importance for policy and action. Statistics about research and development, organized knowledge production, are of

3

comparatively recent origin. The first such regular measurements appeared around the time of World War II. These have been continuously modified and updated to provide increasingly sophisticated treatment of data. The amount of all funds spent for research and development and basic research in 1953 was $5.2 billion; it climbed steadily until by 1970 it reached $26.5 billion. Concomitantly, the total number of working scientists and engineers grew from 557,000 in 1950 to roughly 1,600,000 in 1970.

Obviously there has been a change in the economic structure of massive proportion, linked to the rapid growth and proliferation of technology. Politically, various civil rights movements have pressed for change on behalf of racial and ethnic groups, setting a model expanded to age, sex, and other groupings. All of this is very well known, if not necessarily well understood. Such changes were, not surprisingly, accompanied by far-reaching changes in the symbolic domain of culture. There exists a vast proliferation of symbolism ranging from new languages in mathematics and computer science to diverse bodies of knowledge, variegated modes of artistic expression, and a colorful kaleidoscope of folk culture and styles of life. The often-mentioned *knowledge explosion* is a central component of cultural diversity, differentiation, and growth.

Under these circumstances, many authors have expressed the feeling that there exists now a new societal order or historical era or culture pattern, which the United States has begun to enter. Few observers deny the existence of massive continuities from the past. Yet there seems to be a great deal of substantial agreement that something new has appeared on the scene. The new social order has been referred to as "cybernetic" (Bell, 1967 and 1968), "technotronic" (Brzezinski, 1968), "programmed" (Touraine, 1972), "post-industrial" (Touraine, 1972; Bell, 1973), "knowledgeable" (Lane, 1966), "active" (Etzioni, 1967 and 1968), and "post-modern" (Kavolis, 1970 and 1974; Bell, 1976; J. Marx, 1976)—to mention just a few of the labels that have noted the emergence of the new pattern. This diversity of labels conceals a suprising degree of consensus about the substantive phenomenon to which they refer. For example, Touraine argues (1972:3):

A new type of society is now being formed. These new societies can be labeled post-industrial to stress how different they are from the industrial societies that preceded them. . . . They may also be called technotronic because of the power that dominates them. Or one can call them programmed societies to define them according to the nature of their production methods and economic organization.

and Brzezinski observes that (1967: 18–21):

America is in the midst of a transition that is both unique and baffling. . . . Ceasing to be an industrial society, it is becoming shaped to an ever increasing extent by technology and electronics and thus becoming the first *technotronic society*.

Most social analysts emphasize the critical importance of the technological applications of theoretical knowledge to the productive sector, the shift from the manufacturing economy to a service economy, and also the decreasing economic or class-based conflict between capital and labor or bourgeosie and proletariat. In general, the interpreters of the emerging society have focused more on structural and institutional development or on demographic and ecological changes than on cultural meaning systems. Exceptions to this are the work of Kavolis (1970 and 1974), that of Berger, Berger, and Kellner (1973), and most importantly Bell (1973 and 1976). In these interpretations of recent history we do receive important insights into shifts in the systems of cultural meaning.

Disagreement arises among the analysts of this social change on the question of whether major traditional structural divisions of society will remain and just how they might effect the foci of social conflict. Many believe that fundamental structural divisions and conflicts, based on class ownership and control of the means of production, are diminishing. Toffler (1970), Brzezinski (1967), Aron (1968), Bell (1967, 1968, and 1973), Lane (1966), and Etzioni (1967 and 1968) are among these. One thesis is that regulated, organized competition softened by multiple memberships and affiliations with cross-cutting ties should moderate the forms of conflict in the emerging society. Many of these interpreters assume that there would be increasing public acceptance of rational control by technically competent elites, as well as increasing adoption of the attitudes and common sense of professionals among the middle class. Some feel that scientists, professionals, experts, and other specialists in the ruling elite might be able to chart a relatively harmonious course, minimizing the likelihood that pervasive dissent or dissatisfaction could generate extremist popular ideologies capable of mobilizing large-scale loyalties to movements that would attempt to transform society radically. Others are inclined to argue that the very preoccupation of knowledge-based elites with ideas and their production would destabilize the legitimacy of the political framework and make it subject to constant critique. Lipset argues that knowledge-based elites would work ceaselessly to delegitimize all other elites, while at the same time presenting themselves as a downtrodden group.

Other analyses of contemporary trends, particularly those of Touraine (1972)—while acknowledging that the traditional class conflicts between capital and labor, and the traditional economic bases of domination are losing the central importance of earlier eras—emphasize new cleavages and conflicts emerging between those who control the institutions of decision making, societal integration, and symbol manipulation on the one hand, and those who have been reduced to a condition of dependence, alienation, and powerlessness on the other. Touraine (1972:9) suggests the term "dependent participation" to refer to the alienated masses in contemporary society. Such a neo-Marxian perspective conceives of a new dominant elite or class defined by technical competence and theoretical, scientific knowledge. Neo-Marxian analyses typically anticipate that the new subordinated and alienated class will ultimately revolt against this dependence on a dominating technical elite and embark on an autonomous course that will endow social existence with unprecedented meaning and fulfillment. Understandably, such perspectives are often ambiguous as to the social sources of this new structurally defined group that would provide the dynamic for the creation of a new social order. For example, Touraine indicates that (1972:18):

> . . . sensibility to the new themes of social conflict was not most pronounced in the most highly organized sectors of the working class . . . [rather] . . . the most radical and creative movements appeared in the economically advanced groups, the research agencies, the technicians with skills but not authority, and, of course, in the university community.

In spite of such divergent concerns and disagreements among the perspectives, there is agreement about a vaguely and often imprecisely conceived set of changes that are transforming an advanced industrial order into a qualitatively different type of society. It is less than astounding that there is little agreement on the precise criteria distinguishing the new from the old. Emerging social and cultural trends must, in spite of all social scientific sophistication, remain bewildering to those imbedded in them.

We will not attempt to deal with the full range of issues that this cursory look at history and some of its interpreters has raised. Although ours is a narrower focus, it does require some further observations on certain contemporary trends.

THE POST-MODERN PATTERN

Certain major trends have survived the earlier era of industrialization. They include the marked increase in societal complexity through a much greater division of labor in which many occupations are becoming more specialized and technical. One might argue that the early trend toward urbanization finds its continuation in the growth of suburbs and the great metropolitan complexes. There has been a continuous growth in the power of society over nature as indicated in the consumption of energy. Automation has increasingly limited the role of blue-collar industrial labor and simple clerical jobs. Problems of organization and service appear more salient in this kind of social order than those of production, even though production obviously remains important. All of these could be described as continuing trends that existed prior to World War II. If only such qualitative changes could be noted, we would be puzzled by so many observers proclaiming the existence of a new era.

But there are other changes of more profound significance. There have been marked advances in our ability to handle information of all kinds. Much of this has to do with the proliferation of modes of symbolism and symbol manipulation. Much of it is tied to the advances in electronic media and communication techniques, joined with the widespread availability of inexpensive receivers. These advances have changed the nature and the meaning of mass participation. Technological forms that are appearing on the horizon give promise of even more profound impacts. Advanced transportation facilities transform the traditional conceptions of distance, space and time, making possible—together with the electronic communication media—the emergence of a world community (but not necessarily a conflict-free one). Scientific and technical advances mercilessly render existing knowledge, techniques, and practices obsolete. Knowledge workers in the professional, academic, and technical fields, traditionally thinking of themselves as autonomous professionals and some as intellectuals, experience increasing bureaucratization and controls. These phenomena appear to be changes in configuration and structure, not merely continuations of trends.

Several descriptions of the new kind of society that have appeared in the recent literature of social science merit further scrutiny. Lane (1966:650) captures many of the consequences of the heretofore unprecedented institutionalization of applied, technical knowledge and professional expertise in his definitional statement about the "knowledgeable society":

> Thus, a knowledgeable society would be one where there is much knowledge, and where many people go about the business of knowing in a proper fashion. As a first approximation to a definition, a knowledgeable society is one in which, more than in other societies, its members: (a) inquire into the basis of their beliefs about man, nature, and society; (b) are guided (perhaps unconsciously) by objective standards of veridical truth, and, at the upper levels of education, follow scientific rules of evidence and inference in inquiry; (c) devote considerable resources to this inquiry and thus have a large store of knowledge; (d) collect, organize, and interpret their knowledge in a constant effort to extract further meaning from it for the purposes at hand; (e) employ this knowledge to illuminate (and perhaps modify) their values and goals as well as to advance them. Just as the "democratic society" has a foundation in governmental and interpersonal relations, and the "affluent society" a foundation in economics, so the knowledgeable society has its roots in epistemology and the logic of inquiry.

Citing Machlup (1962), Lane points to increases in personnel and expenditures on knowledge, and concludes that knowledge occupations—producers as well as distributors of knowledge—have grown more rapidly than others. Moreover, the knowledgeable society encourages and rewards men of knowledge far more than it does men of practical business affairs (Lane, 1966:652). Lane's (ibid.: 659–660) further elaboration of the knowledgeable society gives considerable emphasis to the role of technical knowledge applied by problem-oriented professionals who are able to time their specialized technical expertise to public policy and social welfare administration:

> The knowledge society is marked by a relatively greater stress on the use of information veridically, relying on its truth value and not on any adventitious defense, popularity, or reinforcement value. This should be associated with a decline in dogmatic thinking. The decline of dogmatism implies the decline of ideology, in the narrow sense of the term used here. . . . Under the pressure of economic and social knowledge, a growing body of research, and the codified experience of society, ideological argument tends to give way to technical argument, apparently to the disappointment of some.

Whether any particular society has already become fully knowledgeable in character is a question Lane (ibid.:653) does not explicitly answer, but the impression received is that, in his opinion, the United States is not far from this condition:

> The prodigious and increasing resources poured into research, the large and increasing numbers of trained people working on various natural and social "problems," and the expanding productivity resulting from this work, is, at least in size, a new factor in social and . . . in political life. This "second scientific revolution," as it is sometimes called, reflects both a new appreciation of the role of scientific knowledge, and a new merger of western organizational and scientific skill.

As this book will show, matters are a great deal more complex than that.

Etzioni (1967, 1968) envisions essentially similar developments in his characterization of the "active society," which is master of itself and which realizes to the fullest possible extent the goals it has set for itself. This mastery, in turn, presupposes control of knowledge as an essential element in the impending society that Etzioni (1967, 1968) anticipates. He does not portray as clear a division between the producers of knowledge and the decision-making elites as do most authors. On the contrary, he suggests (Etzioni, 1967: 176) that the producers of knowledge play an active role informing the judgments of decision makers, and thus in fact guide societal action and public policy. Etzioni's analysis rests very heavily on the distinction between "stable" and "transforming" knowledge: stable knowledge elaborates what is already known and takes the basic framework of knowledge itself for granted whereas transforming knowledge is concerned with exploring potential challenges to the basic assumptions of a cultural and political system. With this distinction, Etzioni's formulation begins to describe a society not radically different from earlier states of affairs. He goes on to maintain that decision-making elites generally tend to prefer stable over transforming knowledge as well as tend to seek closure on basic knowledge assumptions. A central implication of Etzioni's analysis (1968:213) is that the emerging society may be more concerned with the uses of knowledge than with its production. If such is the case, it would mean that the producers of bodies of knowledge will have more difficulty in establishing their credibility, legitimacy, and worthiness for support than the users and appliers of knowledge. Moreover, Etzioni suggests that the knowledge used, applied, and implemented by societal actors commonly mixes factual information with evaluative interpretation. This would mean that the public would press knowledge producers into applied stances. Knowledge producers, pressed to maintain credibility, would have to convince skeptical publics and political decision makers of the vital functions and applicability of their work. In short, Etzioni's (1967 and 1968) image of the active society involves a state of affairs that is less supportive of knowledge production as such than of the systematic application and utilization of knowledge.

Anderson and Moore make a very dramatic observation when they argue that the increasing reliance of modern society on knowledge production and changing technology will create an emphasis on learning rather than on performance. Since the 1940s, knowledge and technology growth have accelerated. The increase in this process is so significant that it amounts to a qualitative difference. They argue that the "primitive" period of mankind's technological history ended about 1940, with the modern period beginning since then. Their technological-function graphs, on a time scale of 10,000 years, show curves for these functions moving sharply upward in about the 1940's.

They wish to draw particular attention to what they think is the most important consequence of the change.

We think that one important result of this technological leap is that we are in transition from what we have called a "performance" society to a "learning" society. In a performance society it is reasonable to assume that one will practice in adulthood skills which were acquired in youth. That, of course, has been the traditional educational pattern for human beings, and it is reflected in our linguistic connections. We say that a medical student, for instance, learns medicine and the doctor practices it. There is also the practice of law, and in general adults have been the practitioners of skills which they learned as apprentices. In contrast, in a learning society, it is not reasonable to assume that one will practice in adulthood the skills which were acquired as a youth. Instead, we can expect to have several distinct careers within the course of one lifetime. Or, if we stay within one occupational field, it can be taken for granted that it will be fundamentally transformed several times. In a learning society, education is a continuous process—learning must go on and on and on. Anyone who either stops or is somehow prevented from further learning is reduced thereby to the status of an impotent bystander.

(Anderson and Moore, 1969)

In this imagery a large cultural change has important consequences for social structure, and indeed personal identity. It follows from the Moore and Anderson argument that occupational identity, which is so centrally important to industrial society, may be more demanding to maintain in the rapidly changing cultural environment of the learning society.

The best known characterization of the kind of society that is emerging around us today is Daniel Bell's (1968, 1973, and 1976) idea of the "post-industrial" society. In discussing his formulation, Bell notes (1968:158):

The concept of a post-industrial society deals primarily with long-run structural changes in society. It is not, nor can it be, a comprehensive model of the complete society; it does not deal with basic changes in values (such as the hedonism which now legitimates the spending patterns of an affluent society); it can say little about the nature of political crises . . . it cannot assess the quality of the national will . . . by positing certain fundamental shifts in the bases of class positions and modes of access to places in the society. . . .

The society that Bell describes is based on a service economy, and is composed of predominantly white-collar workers, although the society is clearly dominated by the professional class. The most striking of the five dimensions of a post-industrial society that Bell (1976: 198) indicated concerns ". . . the centrality of theoretical knowledge as the source of innovation and policy analysis in the society." Post-industrialism means, for Bell, the preeminence of theoretical knowledge that is characterized by a "new fusion between science and innovation" (Bell, 1968:182). Bell believes that it is precisely the altered awareness of the nature of innovation that will increasingly make theoretical knowledge so central in the emerging post-industrial society. From this, it follows that the chief resource of any post-industrial society is its highly educated, specialized, technical-scientific and professional personnel. Along these lines, Bell predicted that by 1975, 15 percent of the labor force in the United States would be engaged in professional and technical work and that this figure could easily rise to 25 percent by the year 2000, when more than 75 percent of all eighteen- to twenty-five-year olds will have attended college.

Bell's early (1967 and 1968) analyses presented a sharper, more delimited, and empirically rigorous characterization of an emerging societal configuration than his more recent (1973 and 1976) works. Nevertheless, Bell (1976:15) does make it clear that the concept of ". . . the post-industrial society centers on the technology, the kinds of work people do . . . and the organization of knowledge."

For Bell, then, post-industrial society is defined by certain relationships between scientific and technical knowledge and the economic system. Indeed, the level of technology is seen by him as the most critical element; but in his latest work he also focuses critically on the destabilizing role of modernist expressive culture. Whatever the merits of Bell's specific argument on this point, it is interesting to note that all of these diverse characterizations of the emerging social pattern emphasize the increasingly important role of the specialized, technical knowledge. As a consequence, they present a picture of a social structure in which knowledge workers such as scientists and professionals occupy a particularly important place. The conception of the cultural system that accompanies these notions is one of rapid change, diversity, and specialization. Indeed, the differentiation of the social bases for knowledge supports increasingly divergent culture products.

The preceding observations raise more issues and questions than we can deal with in this book. When we adopt the language usage of referring to the new pattern of culture visible in the United States as post-modern we do it not without misgivings. Whether this pattern should be seen primarily as an outgrowth of uniquely American historical circumstances, or of the dynamics of the capitalist system, or whether it is an evolutionary stage remain background puzzles. We have no reason to believe that societal evolution follows one single path. Indeed, differing civilizational codes and cultural rationales demonstrate considerable long-range stability, influencing the shaping of modernity in their respective parts of the world in quite distinctive ways. In American history, we can see such a thematic invariance. At the same time, it does not seem plausible to us to claim that everything about the post-modern pattern is a uniquely American or capitalist phenomenon. There are compelling and universal consequences from an emphasis on explicit and organized knowledge use in economic, social, and political life. Just how historical idiosyncracies, evolutions, and other elements are to be analyzed apart we do not know. A massive effort and comparative work would be necessary for this purpose.

Our focus is the knowledge system itself. We offer an exploration and what we believe to be an important analytical sketch. Most of our illustrations are drawn from American society even though the effort remains primarily conceptual, not historical. In doing this, we wish to focus attention on the general sociological consideration of the social system for knowledge of which the contemporary United States offers a single, but a highly instructive, instance.

The knowledge system can be seen as social arrangements clustering around the processes of knowledge production, organization and storage, distribution, and use. Such distinctions have become quite commonplace in the sociology of the professions. The literature on the subject is enormous: Moore (1970) recently listed more than 850 references to the literature on professions alone. E. C. Hughes

(1958:139-144) distinguished among three occupational models: science, profession, and business. Science is concerned with the discovery and systematization of knowledge; the professions represent the giving of esoteric services to clients who, as laymen, cannot be expected to handle highly technical problems by themselves. The clients are further perceived as incapable of evaluating the work of the professional, whose practice is based on some body of applied theory. Freidson (1970) used a similar distinction between, on the one hand, the scholarly and learned professions that create and elaborate the formal knowledge of a civilization and, on the other hand, the practicing or consulting professions that apply that knowledge to everyday life. Ben-David (1964) used an analogous distinction between the creators and the reproducers of knowledge.

It should be quite clear that no occupation is composed entirely of people whose work represents only one or the other of these professional roles. Yet there is a tendency for such separation. We can, indeed, distinguish between knowledge production roles and other knowledge structures. The conception of the knowledge user is not, of course, coextensive with that of the practicing professional. Business, the formation of policy, social and political movements, all are among the users of organized knowledge. We will deal quite explicitly with some patterns in that domain.

It is particularly interesting that in the post-modern social system we find an unusual diversity of roles and institutions that deal with the organization, storage, and distribution of knowledge. For knowledge that has been recently produced or discovered to become accessible to professional practitioners, it must be organized into coherent and consistent bodies of knowledge. One example for this kind of symbolic organization can be found in professional textbooks, monographs, and manuals.

These illustrations unquestionably do not exhaust a description of the new pattern. We are suggesting, however, that the emergence of mediating roles, statuses, and functions that link knowledge production and use is an inevitable consequence of the increasing formalization and differentiation of the knowledge system. Much of this follows from the quest for explicit and systematic knowledge application. This particular aspect is still, in spite of recently rising interest, a somewhat neglected aspect in the sociology of knowledge. We will emphasize it particularly.

The idea of post-modern society is that of a relatively self-conscious, reflective social system, which some view as an object capable of deliberate change through the systematic use of organized knowledge. Thus, whereas the idea of modernization generally refers to the institutional concomitance of technologically induced economic growth and the social transformations that accompany it (see for example Berger, Berger, and Kellner, 1973), the idea of post-modern society stresses the organized application of knowledge. It is in this sense that the term post-modern society is synonymous with knowledge-based society (Holzner, 1972). The expectation that organized knowledge in the sciences and technological and professional fields is to be a major factor in production, policy forming, and indeed in the quality of life experience by citizens is virtually taken for granted today. Local as well as national governments are compelled to formulate and implement not only general science policies, but also policies and programs for the concrete application of knowledge. The federal government in the United States, for example, has designed intricate policies for knowledge production and use in such diverse areas as space research, military technology, public health, education, and many other fields. In fact, citizens certainly feel that their governments have a responsibility for the production and availability of potentially needed knowledge in whatever area problems may arise.

This circumstance illustrates one of the conditions for such explicit knowledge use; namely, trust in bodies of knowledge. By this we mean an attitude that assumes one can safely attempt to calculate risks of taking action on their basis. Just as knowledge reduces uncertainty by creating ordered information, trust in knowledge reduces complexity. Beyond this underlying trust in the reliability of existing bodies of knowledge, post-modern societies have a deep faith in the efficacy of organized knowledge and in the ability to produce it as needed through the application of an organized technical effort.

We do not have to search far for illustrations of this historically unique conviction that urgent societal problems are basically amenable to resolution through systematic knowledge use. There is an impressive faith in the producibility of relevant knowledge. The apparent planfulness of the program to land a man on the moon probably serves as an important popular paradigm for this belief.

This very notion of knowledge being capable of production by plan, or following very detailed schedules, is certainly relatively novel and without historical precedent. The public attitude underlying the conception that it is a governmental responsibility to budget and to plan for knowledge production and use is rarely discussed or thought about as the historic novelty it is. The rapid growth of knowledge-producing organizations has often been noted, and the extended reliance on technical, professional expertise for dealing with all kinds of action problems hardly requires documentation. But the massive effects of the qualitative changes in the nature and structure of post-modern societal communities that result from the large increases in their knowledge-based sector have not as yet received enough systematic and analytic attention.

While many writers have stressed the increased importance of the production or discovery of specialized knowledge, few have systematically and seriously addressed themselves to the question of how the information that is produced ". . . gets established as recognized knowledge, and how its development and utilization become organized, evaluated, and controlled" (Freidson, 1970:28). In other words, none of these previous treatments give adequate attention to the unusual diversity of role statuses and institutions for organizing, distributing, and storing knowledge. These features are, we believe, a major and central aspect of post-modern culture. For it is only with the emergence of structurally differentiated, functionally specific role statuses and positions for organizing, distributing-transmitting, and storing-retrieving specialized, technical knowledge that knowledge application is institutionally linked to knowledge production-creation. It is in this manner that the specific post-modern knowledge system comes into being.

Both the emphasis on the organization, distribution-transmission, and storage-retrieval of knowledge as well as on the institutionalization of these processes distinguish the present analysis from previous discussions of the nature and significance of knowledge production-producers and knowledge application-users in contemporary society. Yet, we believe these assumptions underlying the present formulation to represent plausible, observable, and defensible points of departure.

Of course, knowledge creation and production have been present, at least in unorganized, uninstitutionalized fashion, as long as men have lived in groups. Certainly the invention, discovery, or production of new knowledge has been a feature of social life since the scientific revolution. Analogously, although men have informally used or applied knowledge since the time when the species first developed the ability to symbolize, even the formal, organized, and institutionalized application of specialized knowledge can be traced back over many centuries. In short, knowledge producers and the discovery of new specialized, technical information, on one hand, and knowledge users or the systematic utilization of that information and its application to concrete human problems, on the other, have been with us for some time. Structurally, as well as functionally, neither the role statuses nor the processes are historically unique. The special feature of the current situation, in addition to the quantitative increase in knowledge, is the emergence of role statuses and processes self-consciously and intentionally designed to link the production-discovery of new knowledge with its application-use in a coherent, integrated, purposive, and pervasive knowledge system.

In a recent article dealing with "new networks of knowledge and information in post-industrial society," the leading prophet of the sociocultural future, Daniel Bell, predicts that:

> The major structural change—and the great economic dislocation—is in the area of transmission. The problem arises because of the blurring of the technological and legal differences between telecommunications and teleprocessing, the latter being not only the transmission of computerized data information, but the processing of news and library materials as well. Technologically, telecommunications and teleprocessing are emerging into a mode which Anthony Oettinger has called *compunications.* As computers become used increasingly as switching devices in communication networks, while electronic communications facilities become intrinsic elements in computer data processing services, the distinction between processing and communication becomes indistinguishable.
>
> (Daniel Bell, *Encounter*, 1977:12)

What Bell refers to as transmission corresponds roughly to the functionally related but structurally differentiated process of organizing, distributing, and storing-retrieving knowledge and information. And as Bell (ibid.:14) is quick to realize and explicitly acknowledge, the developments and then increasing fusion of communication technologies for organizing, distributing-disseminating, and storing-retrieving

knowledge and information raise questions that are not only technological and economic, but even more importantly, political:

> Information is power. Control over communication services is a
> source of power, and access to communication is a condition of
> freedom. There are legal questions that derive directly from this.

In short, the development of institutionalized roles, positions, and
processes for organizing, distributing, and disseminating, as well as
storing and retrieving specialized, technical knowledge not only
generates an historically unique cultural system, but profoundly influences the critical issues that must be confronted in the political arena.
Thus, it is almost impossible to overemphasize the importance of
those previously neglected role statuses and processes that organize,
distribute-disseminate, and store-retrieve specialized, technical knowledge as well as link knowledge production-producers to knowledge
application-users, thereby creating a coherent post-modern knowledge
system.

   Not surprisingly, these developments go hand in hand with
heightened public ambivalence about experts, scientists, professionals,
and academics. Their authority in relation to practical affairs, programs, and problems is particularly under challenge. Nearly everyone
has had a few frustrating encounters with the representatives of specialized, technical expertise, be they in medicine, law, television re-
pair, or automobile service. Problems associated with the expert's
authority and competence in problem solving as well as issues of
public accountability are being raised increasingly. Complaints about
the overspecialization of technical experts are being voiced as are
complaints that experts express themselves in language too convoluted
for ordinary people. These complaints often come from people who
are, in some other respects, experts themselves.

   Even more important is the public unease that the aggregate
effort of sophisticated corps of experts often proves ineffective in
the stubborn and complex face of reality and fails to produce the
desired results. Certainly, America's experience with the social programs of the 1960's was, at best, a mixed one. Frustration predomi-
nated, often resulting from the overconfidence in the self-presentation
of presumed experts. The effort to aggregate decisions through the
inputs of different kinds of technical experts and specialized compe-

tencies sometimes even led to absurdly tragic blunders, as in the Vietnam War. In fact, as many have noted, the very rise of specialized professionalism and technical, bureaucratic expertise in post-modern society has been accompanied by frustration, disenchantment, alienation, and even by movements espousing mysticism and irrationality.

The pervasive characteristics of the post-modern pattern also can be detected in the forms of social movements. Movements are deliberate mobilizations for or against change; they always are connected with conceptions of possibilities alternative to the given reality. The major patterns of social movements in industrial and industrializing societies were large scale mobilizations for institutional change. Political parties, military organizations, or revolutionary cadres were and are the prototype of this pattern. Movements dealing with grievances against institutional patterns, the distribution of political power and economic reward, and the like persist in vitality. But the older form is being supplemented by cultural movements. Contemporary movements may be thought of as media contexts, channels, and mechanisms for the ideological construction of models of identity. They become legitimate contexts for intentional adult resocialization. One may speak of movements for identity transformation and reconstruction. The movement for black power and identity, contemporary feminism, and the various student-youth movements are illustrations of this phenomenon (J. Marx and Holzner, 1975).

ON COMMON SENSE

Superficially, common sense is what the differentiated, pluralistic, post-modern society appears to lack. Certainly, surfeits of information, differentiated rationales, and decision criteria used by experts, and sensory overload do in fact erode the confidence and certainty a practical person may have in the adequacy of his common stock of traditional knowledge. It may at any moment be superceded by either new knowledge or new fashions.

Common sense in this kind of highly differentiated, self-reflective, knowledge-oriented society must necessarily take a form that is different from the pattern it took in a society in which most knowledge at hand was traditionally transmitted, learned through apprenticeship, or generated in its use. The common sense of post-modern, knowledge-based society requires acceptance of varied, differentiated frames of reference and confidence that rules exist or can be discovered for their translation into each other. It also requires a fairly complex understanding of, and faith in, rationality as an encompassing code that admits of differential specifications and modes of expression, but also provides for their ultimate mutual intelligibility. In other words, the common sense of highly advanced societies rests on the assumption that differentiated, specialized bodies of knowledge are in principle accessible and usable by anyone who is willing to undergo the necessary training to grasp them. Furthermore, common sense—this body of shared understandings and symbols—makes specialized knowledge and expertise accessible by defining general routes for knowledge search and conditions for use. Finally, it provides a code for responsible conduct in which the use of expertise and knowledge plays a special role.

The notion of common sense that has found its way into scientific as well as philosophical discourse has been rather simple. It is a notion of shared substantive understandings and maxims of living. Thus, common sense is attributed to individual actors; it is rarely conceived of as a collective achievement that bestows certain qualities on events and situations. What distinguishes common sense as a mode of *seeing* is largely a simple acceptance of the world, of its objects, and its processes as being precisely what they seem to be. It is joined to a pragmatic motive, the wish to act upon the world so as to bend it to one's practical purposes, to master it, or adjust to it (Schutz, 1962; Geertz, 1966). But the common sense world of everyday life that is both the established context and given object of our actions is also a shared cultural product framed in terms of symbolic conceptions of stubborn fact. It is structured around the domain of paramount reality.

As Geertz notes (1973:5), this last point undermines "the unspoken premise from which common sense draws its authority—that it presents reality neat—." Geertz goes on to observe:

If common sense is as much an interpretation of the immediacies of experience, a gloss on them, as are myth, painting, epistemology, or whatever, then it is like them, historically constructed and, like them, subjected to historically defined standards of judgment. It can be questioned, disputed, affirmed, developed, formalized, contemplated, even taught, and it can vary dramatically from one people to the next. It is, in short, a cultural system, though not usually a very tightly integrated one, and it rests on the same basis that any other such system rests; the conviction by those whose possession it is of its value and validity.*

Now insofar as common sense is a cultural system, a particular perspective or way of looking at the world that rests on a shared symbolic template, it is possible to treat it as analogous to science, religion, and ideology as cultural systems. Therefore, it is possible, in analogy to Geertz's (1966:90) conception of "Religion as a Cultural System" to define common sense as a symbol system that acts to establish pervasive and long-lasting moods and motivations in people by formulating conceptions of everyday reality—the world, its objects, and its processes—and clothing these conceptions with such an aura of ineluctable factuality that the modes and motivations seem uniquely appropriate and necessary. The distinguishing feature of the conceptions of everyday reality associated with common sense is that the world is accepted as given—the world is what one takes it to be—and as an appropriate object of deliberate, purposive attempts to transform, utilize, or adjust to it.

Science, as the distinctive and pivotal cultural symbol system in post-modern society, contrasts with the common sense perspective, while it is embedded in it. In the scientific perspective, with its special disciplines of organized skepticism to detect error in empirical observation or rational calculation, it is precisely the givenness, the simple and straightforward acceptance of the world that disappears. Science as a cultural system institutionalizes deliberate doubt and systematic inquiry, it suspends the pragmatic motive in favor of disinterested observation and skepticism. Science attempts to analyze the world in terms of formal concepts whose relationship to the more implicit and informal—although no less problematic—conceptions of common sense becomes increasingly tenuous. Both science and common sense, however, share the characteristics of all cultural symbol systems: they are socially constructed meaning systems that provide models and interpretations for events which would otherwise be problematic or meaningless. In other respects, however, they seem antithetical. Science is an outgrowth of sociocultural differentiation: it is practiced

*Reprinted by permission of the author.

by specialists who are presumed to have special knowledge skills acquired through extensive and unusual training procedures. There results the formation of a community united in the idea that epistemic criteria of judgment are supreme. Thus, the standards and knowledge criteria of scientists become different from those employed by "everyone" in dealing with the realities of everyday life. In contrast, one of the hallmarks of common sense is what Geertz calls its "accessibleness":

> Accessibleness is simply the assumption, in fact the insistence, that any person with their faculties reasonably intact can grasp common sense conclusions, and indeed, once they are unequivocally enough stated will not only grasp but embrace them . . . there are really no acknowledged specialists in common sense. Everyone thinks he's an expert. Being common, common sense is open to all, the general property of at least, as we would put it, all solid citizens. Indeed, its tone is even anti-expert, if not anti-intellectual: we reject, and so far as I can see, do other peoples, any explicit claim to special powers in this regard. There is no esoteric knowledge, no special technique or peculiar giftedness, and little or no specialized training—only what we rather redundantly call experience and rather mysteriously call maturity—involved. Common sense, to put it another way, represents the world as a familiar world, one everyone can, and should recognize, and within which everyone stands, or should, on his own feet.
>
> *(Geertz, 1973:29–30)*

The post-modern society is a special social pattern in that it is specifically related to the dynamics of explicit knowledge in science, technology, and professional expertise. Yet it would not be correct to argue that the bodies of scientific, technological, or professional knowledge permeate this society. Indeed, most citizens of such societies are not involved in the cultural system of science in any serious way. And yet the impact of science and technology on other cultural domains, and especially on common sense, has been massive and profound. This transformation is of particular importance. We will show later the social embeddedness of specialized knowledge structures; here our concern is to point to the fact that the cultural system of common sense in post-modern society has undergone alterations accommodated to the peculiar exigencies of the knowledge system.

Again, Geertz captures this point (1973:22):

> The development of modern science has had a profound effect—though perhaps not so profound as sometimes imagined—upon Western common sense views. Whether, as I rather doubt, the plain man has become a genuine Copernican or not (to me, the sun still rises and shines upon the earth), he has surely been brought round, and quite recently, to a version of the germ theory of disease. The merest television commercial demonstrates that. But, as the merest television commercial also demonstrates, it is a bit of common sense, not as an articulated scientific theory, that he believes it. He may have moved beyond "feed a cold and starve a fever," but only to "brush your teeth twice a day and see your dentist twice a year."

In other words, the consequence of modern science and technology has not been to overwhelm common sense views and render them insignificant, but to modify them somewhat and increase their cultural significance to the point where they represent the unique and distinguishing characteristic of post-modern society. Other historical societies have spawned specialized cultural domains in which esoteric bodies of knowledge were produced and specialized technical skills employed by practitioners who came to form an autonomous epistemic community—witchcraft, alchemy, astrology, and acupuncture are but a few of the analogous cultural domains one can suggest. What is perhaps unique about modern science is the impact it has had on other cultural meaning systems—religion, art, ideology, and especially common sense—by serving as a more general code or paradigm for all bodies of knowledge in all cultural domains. The demonstrable consequence of scientific and technological rationality produced the rationalization of all major cultural symbol systems and the faith in socially organized, systematic rationality that permeates common sense views in post-modern society.

Perhaps the most characteristic manifestation of this belief in specialized, institutionalized knowledge and expertise is that it prevails even in instances where action must be taken on the basis of knowledge that departs far from intuition, daily experience, and tradition. An example is the flight of Apollo VIII, the first translunar manned space voyage, during which the command module left earth's orbit, traveled to the moon, and circled it for three days before leaving the moon's orbit and returning to earth. For the average person, with only a dim understanding of gravitational fields, the idea that an object fired into space on a calculated trajectory such that it would circle the moon for a while and then return to a particular place on earth, carrying three men, staggers the imagination. Yet the important point for us here is not so much that this feat was carried out successfully, but that nearly all Americans, after being informed of the plans confidently expected and believed that it would be. It is in this faith in specialized, institutionalized bodies of knowledge and expertise that we find one of the important links that keeps post-modern complexity grounded in a new fabric of shared, common sense trust.

Let us consider only one further illustration of the kinds of transformations that we believe have occurred in common sense typical of the post-modern pattern. Consider the change that popular, common sense views of disease and illness have undergone over the past century in relation to scientific medicine and the medical profession. Scientific medicine represents the application of knowledge drawn from the natural sciences to concrete problems through the specialized expertise of a highly disciplined community of practitioners. Over the past 100 years, popular belief in the validity and efficacy of scientific knowledge has grown to the point at which the organized, systematic application of that knowledge by a professional group has come to be regarded unproblematically as the sensible, rational, obvious way to alleviate the suffering of illness. The decline of proverbial folk culture in the curing of illness—while by no means complete—is certainly impressive. The most significant change that has taken place involves the incorporation of a professional medical-disease model of illness into popular assumptions and symbols that structure and give meaning to behavior and experience.

The medical-disease model of illness is a specialized conceptual framework that evolved in the practice of physical medicine and subsequently became incorporated into popular beliefs and collective common sense assumptions. Such popular beliefs involved common sense assumptions concerning etiology, pathology, therapy, and professional authority. At the common sense level, it is assumed that symptom clusters can be diagnosed and that they form discrete disease entities or pathological processes underlying these symptoms.

It is assumed that the underlying illness process produces physical or mental disease, that internal processes are responsible for an individual losing control of physical well being or social behavior, and that each underlying disease produces a distinct symptomatology. The etiological views embedded in this common sense conception assume that a pernicious agent is the causal factor in a sequence leading to mental or physical disease. The therapeutic convictions involve the assumption, especially critical for the maintenance of the model, that pathological processes within the organism can be effectively treated by a physician, a professional medical scientist specializing in the application of technical expertise to problems of human health. The behavioral posture of a person seeking medical advice as patient is well established. At the same time, common sense does not value or encourage people to acquire sufficient substantive knowledge themselves to treat their own diseases. On the contrary, common sense maxims emphasize the importance of seeking help from qualified practitioners.

Indeed, the cultural model of disease and its appropriate treatment has become an important source for orienting oneself also to other issues. The use of such a conception for the diagnosis of social problems has become rather pronounced. More generally, the model of professionalism has influenced reform movements. Moynihan (1968) argued that the professionalization of the middle class culminated in the "professionalization of reform" typified by community action programs of the war on proverty in the mid-1960's.

To sum up, we have tried to show that the unprecedently rapid and extensive sociocultural differentiation that characterizes post-modern society has had its impact on the cultural system called common sense. Instead of emphasizing primarily substantive maxims of daily living (the proverbial guidance of folk wisdom), common sense now provides an orientation to sources of needed information and procedures for gaining access to specialized bodies of knowledge and their expert practitioners. Thus, the social function of common sense in the reduction of complexity and uncertainty has not changed, but the emphasis on stable, substantive capsules of folk knowledge has been supplemented by a means-oriented attitude to accessing information that is known to be socially stored. It is important that this has required an acceptance of divergent frames of reference and rationales for action that provide an impressive, if at times fragile, shared basis of trust in knowledge.

2

The Quest for Knowledge Application

In this chapter, we present an overview of some selected conceptions of how knowledge relates to action, to contexts of its application. The problem, of course, is one of great importance for philosophy and intellectual life. We have chosen certain conceptions that appear to us to illustrate major aspects of the issue that are important for an understanding of its sociological significance. We begin with the emergence of the sharp juxtaposition between knowledge conceived of as important in its own right and the modern view that knowledge always relates to some mode of application.

TWO KINDS OF KNOWLEDGE

An illuminating exploration of two contrasting kinds of knowledge is presented by Hans Jonas in his essay on "The Practical Uses of Theory" (1959). He juxtaposes a quotation from Thomas Aquinas' commentary on Aristotle's "On the Soul" with a sharply contrasting statement from Francis Bacon's "The Great Instoration." Portraying knowledge as noble truth, Aquinas wrote:

All knowledge is obviously good because the good of anything is that which belongs to the fullness of being which all things seek after and desire; and man as man reaches fullness of being through knowledge. Now of good things some are just valuable, namely, those which are useful in view of some end—as we value a good horse because it runs well; thus other good things are also honorable: namely, those that exist for their own sake, for we give honor to ends, not to means. Of the sciences some are practical, others speculative; the difference being that the former are for the sake of some work to be done, while the latter are for their own sake. The speculative sciences are therefore honorable as well as good, but the practical are only valuable.

(Jonas, 1959)

Francis Bacon proposed a utilitarian view of knowledge as a weapon or tool to be applied in the protection of humanity from life's "necessities and miseries."

I would address one general admonition to all: that they consider what are the true ends of knowledge, and that they seek it not either for pleasure of the mind, or for contention, or for superiority to others . . . and for the benefit and use of life, and that they perfect and govern it in charity . . . (from the marriage of the Mind and the Universe) there may spring helps to man, and the line and race of inventions that may in some degree subdue and overcome the necessities and miseries of humanity . . . for the matter in hand is no mere felicity of speculation, but the real business and fortunes of the human race, and all power of operation. . . . And so those twin objects, human knowledge and human power, do really meet in one.

(Jonas, 1959)

Since the time of Francis Bacon's *Declaration of Independence* from the scholastic learning, which believed in a divinely ordered universe, science has followed the path he recommended. Most people have adhered to the belief that "knowledge is power," yet increasing belief in progress through the application of systematic, organized, scientific knowledge had gone hand-in-hand with increasing skepticism about common human reason. The urgent quest to control nature by furthering the understanding of its workings demands precisely such an attitude. From the theory of idols to the psychology and sociology of knowledge, a long tradition of skepticism questions what may be taken for knowledge. Faith in organized scientific knowledge and the efficacy of specialized, technical expertise are accompanied by an ambivalent distrust and skepticism.

This theme relates to the ambivalences of modernity. Their existence and special severity in post-modern society are not a novel discovery. They have often been commented on (see, for example, Aron, 1968; Bendix, 1964; and Berger, Berger, and Kellner, 1973). Much of the systematic reflection on this matter has not dealt with the characteristics of contemporary advanced society nor the consequences knowledge and expertise have in them. It is with respect to these characteristics that one finds forces for both the promise and the predicament of contemporary advanced society, forces that relate directly to the dynamics, motivations, and constraints of knowledge production, dissemination, and use. As we explore these matters, our focus is on the problem of knowledge application. A fruitful understanding of the issues inherent in knowledge application requires a very detailed grasp of specific problems and problem domains as well as knowledge or expertise to be applied to them. Even more important is a global comprehension of the larger societal contexts and cultural configurations in which it all occurs.

THE SPECIAL ROLE OF FACTUAL SOCIAL KNOWLEDGE

Concern with organized knowledge and its application is a product of the vast historical transformation of society, which Max Weber called "rationalization" and which we have come to see as the most significant aspect of the modernization process. Increasing reliance or rationality in handling information and making decisions—and the increments in societal power that result from it—is the strategically crucial aspect of social development. Such reliance on rationality generated its most spectacular results in mathematics and the physical sciences. The gradual construction of an autonomous domain within which systematic rules for empirical inquiry and theory construction prevail became the social and cultural foundation for the explosive growth of scientific knowledge. The instrumental use of this knowledge for economic purposes and for military technology of various kinds has become the paradigm for all knowledge utilization. However, rationalization and institutionalized knowledge production have certainly not been limited to the domain of science. An equally essential rationalization process has occurred in the domain of law and, connected with it, in the domain of public administration.

There is little doubt that the systematic codification of law, both in terms of procedures and with regard to the establishment of legal facts, created a framework of rules within which rational calculations were possible. This provided the essential context for economic rationalization itself. Equally important is the fact that the building of modern states and their complex systems of public administration generated enormous new knowledge needs and strenuous efforts to satisfy them. The growing power interests of the new Western nation-states required military, economic, and administrative information for making governmental decisions. The increasingly sophisticated methods devised for gathering social facts and recording decisions on vitally important population dynamics and economic transactions spun an intricate network of social measurements and measuring devices throughout these emerging modern national societies.

The enormous amount of social knowledge generated in this fashion, though not scientific, certainly is factual. Use of such knowledge, even more than the use of specialized, technological knowledge, has given rise to controversy. Factual social knowledge, more than technical knowledge of the engineering variety, seems closely connected to at least the potential for social control. Indeed, interests in social control often were the motives underlying the collection of such information. Systems of social accounting were often designed to meet military or taxation needs. They provided regularized, bureaucratic methods for collecting factual data about social events and characteristics so that they could be easily systematized and utilized in decision-making. The same facts could become the bases for arguments of social protest and for social reform. The policy significance of the increasing factual knowledge base which post-modern states have constructed about their own societies was and is rather obvious. Therefore, this knowledge itself has become increasingly controversial.

We emphasize the special importance of social knowledge here because of its central relevance to forms of social consciousness and to the reflexive self-understanding of post-modern societies. When, in biblical times, King David instructed the commanders of his army to "go through all the tribes of Israel, from Dan to Beersheba, and number the people, that I may know the number of people," they not only remonstrated with him, but dreadful consequences followed:

> David's heart smote him after he had numbered the people. And David said to the Lord, "I have sinned greatly in what I have done. But now, oh Lord, I pray thee take away the iniquity of thy servant; for I have done very foolishly."

God offered King David the choice between three years of famine, fleeing for three months before his enemies, or three days pestilence in his land. David, incidentally, chose the three days pestilence—apparently as the punishment that would endanger his rule least (2 Samuel 24). Clearly, taking a census of the people of Israel in those days was not a light matter.

In modern states, gathering statistics and conducting censuses have become institutionalized. The United States Constitution requires a census every decade as well as distribution of information about the "state of the union." As products of the enlightenment and revolutionaries in the service of reason, the Constitution's authors recognized the centrality of social accounting and reliable social knowledge for the modern nation-state. The growth of factual social knowledge certainly contributed heavily to the growth of state power. It also contributed to the development of critical, and sometimes revolutionary, social thought. The systematic, scientific study of society as an object and the construction of models of alternative social systems depended on the availability of factual information about social states of affairs. The most recent efforts to improve the information base for social understanding and social action are attempts to construct "social indicators." This social scientific enterprise represents an attempt to construct measurements capable of displaying not only the state of the national economy or of the administrative effectiveness of the government, but also of the psychosocial well-being of the citizens.

The intimate connection between various efforts as knowledge production and the self-conscious construction of (future) post modern societies is apparent in the impact that the development of social indicators has had on the emergence of forecasting the future as a deliberate, rational, and systematic enterprise. The acquisition of reasonably reliable social knowledge about societies has generated more organized, publicly sponsored and approved forecasting efforts by organized groups. Physical and social scientists as well as engineers have undertaken systematic attempts to predict and affect the future. Such attempts, based on systematic, factual social knowledge of present societal states of affairs, illustrate the link between knowledge production efforts and the modification of selected images for the future. They also illustrate the intimate relationship between institutionalized knowledge production and both self-conscious societal transformation and social control.

A special problem exists with regard to the application of social science. This is true because social science—both as a differentiated body of specialized knowledge and technical expertise, on the one hand, and as an intellectual community of professional practitioners, on the other—is part of the social reality it investigates. The embeddedness of knowledge in a social context is particularly and often painfully obvious in all social science work. Moreover, explanations of social states of affairs may influence those very facts by becoming generally diffused cultural models for them.

Social theories can, of course, be learned by the people whose actions they are designed to explain. The history of social thought and, in a lesser way, that of social science abound with examples of such transformations of knowledge through its reification. Certainly, theories of economics have helped to shape economic institutions and provided cultural models for economic conduct. Similarly, psychoanalysis has profoundly altered not only the explanation of human conduct but, among those who have learned and accepted its vocabulary, it has become a grammar of motives. It is in this sense that social knowledge relates very closely to the cultural models for social life. As Geertz has it (1966), descriptive "models of" may become "models for" both personal and collective consciousness.

We have suggested that social scientific knowledge about behavior and/or society generates new conceptual models of social life —new theoretical interpretations or generalizations. Application of that knowledge may produce new templates or models for the patterning of personal and collective consciousness among those who possess, apply, or are affected by that knowledge. Such knowledge is applied to increasing ranges of people and situations; the new models of/for consciousness that derive from it are first disseminated and circulated, then discussed and routinized, and finally taken for granted. In short, they become part of the cultural domain of common sense. These models may in turn become templates for psychosocial characteristics that everyone, or nearly everyone, views as quite natural and obviously inherent in any normal person. Finally, these common sense cultural models of/for individual and collective behavior and consciousness often become the basis of subsequent social scientific knowledge production and application—which initiates a new cycle. The central relationships involved in this process may be presented schematically as follows:

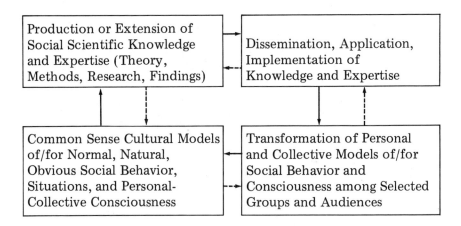

(The solid arrows indicate the primary direction of the relationship that has been described above; the arrows with dashes indicate the dialectical nature of each of these interactions and the fact that the sequence can move in the other direction.)

SOCIAL DIFFERENTIATION AND THE RATIONALES FOR KNOWLEDGE

Inevitably, knowledge change and growth have also meant differentiation, even fragmentation. All systematic knowledge construction in the sciences and professions requires the development of special rationales, perspectives, frames of reference, and specialized languages as well. Every knowledge domain requires certain skills, often of a highly specialized nature, of the practitioner and user of the knowledge. It is precisely this increasing differentiation and specialization, as well as the proliferation of differently structured bodies of knowledge—in the academic disciplines, in the domain of technology and engineering, in the professions, and in the arenas of political and public administration—that has generated the modern problems of knowledge application. This differentiation of specialized knowledge domains and bodies of knowledge is a consequence of the necessity of autonomy for specialized inquiry leading to knowledge production. By autonomy, we merely mean a social condition that permits an inquirer to pursue and attend to the implications of the (any) consistency that inheres in the facts and rationales with which the inquirer works. In the (natural) sciences and in the (established) professions, such autonomy claims have been vigorously and successfully defended. In consequence, conceptions of the unity of all knowledge have been put under severe strain, insofar as they have survived at all. The need for autonomy generates differentiated, specialized bodies of knowledge.

The application of knowledge means its transfer from one specialized domain to another: the problem of knowledge application arises only in those situations in which the knowledge at hand is insufficient to deal with the problems an actor encounters, so that one must turn to knowledge produced by others, often in another domain. (Only where the knowledge at hand is produced in the same framework and context in which action is required can there be no problem of knowledge application.) In modern and especially post-modern societies, therefore, applying knowledge means that it must travel from one cultural domain to another through channels of communication. This circumstance makes the problem of knowledge application so central to understanding modernity and post-modern societies. But the same circumstance also requires an understanding of the fine-grained texture and structure of these societies in order to comprehend the dynamics of knowledge application adequately.

In the highly complex technological societies, for example, one can speak of knowledge supply and knowledge demand, using a marketplace economic analogy. In fact, it is possible to view the problem of knowledge surfeits and deficits as a question regarding the distribution of knowledge as a desired good. The economics of knowledge distribution is spelled out in Fritz Machlup's book *The Production and Distribution of Knowledge in the United States* (1962). Whatever the contemporary appeal of such imagery, this is clearly not the only way in which one can think or in which men have thought about the issue. We now turn to some earlier conceptions of the manner in which the problem of knowledge application was conceived and formulated.

THE PROBLEM OF THEORY AND PRACTICE— SOME HISTORICAL CONCEPTIONS OF KNOWLEDGE APPLICATION

The problem of knowledge application has historically been discussed as the problem of the relation between theory and knowledge in action, often under the heading of the relation between theory and practice. Formulated in these terms, it was a central issue that permeated the body of Enlightenment thought that placed its faith in reason. Reason was understood as the powerful force that transforms society and the entire human enterprise. The light of reason would overcome the darkness of ignorance, prejudice, and evil for the obvious and unambiguous betterment of the human condition. Since that time, however, men have become more skeptical about the utility and validity of reason and of social knowledge. This skepticism and the process of differentiation have made the relation between theory and knowledge in action, the problem of knowledge application, increasingly significant. In this section, we will discuss only a few historical models of the relation between theory and practice, those that point up conspicuous differences in conception and approaches to the issue.

Immanuel Kant addressed the issue of the relation between theory and practice in his essay "About the Commonplace Saying: This May Be Valid in Theory But Is Not Suitable for Practice." He was concerned with the applicability of universal moral theory and his reflections on the matter led him to conclude that anything that is true in theory must be possible in practice. Specifically, Kant concluded (1964:127ff):

A set of even practical principles is called "theory" if these rules, as principles, are thought of with a certain universality, and if a number of conditions are abstracted from them which still necessarily influence their application. Conversely, we call practice not every activity but only that pursuit of a purpose which is thought of as application of certain general principles of procedure. Obviously, there needs to be a connecting and mediating link between theory and practice, no matter how complete the theory, since the concept of understanding which is contained in the rules must be applied by an act of judgment whereby the practitioner distinguishes whether something is the case or not.

It is interesting that Kant distinguished between activity and practice. Practice is disciplined conduct: it is calculated—at least in part—and thus differs in its methodical nature from other kinds of activity. Knowledge is applied to practice in the sense that the theoretical principles of universal validity become the ground for the formulation of the principles of practice. However, there is no way in which practice can be directly deduced from theory. There always remains an element of judgment through which the practitioner "distinguishes whether something is the case or not." This Kantian conception of the relation between theory and practice is one that stresses rational, enlightened practical conduct on the basis of theory. Such conduct requires discipline and must be methodical in nature—not every activity qualifies.

One can certainly find innumerable instances of this conception in the contemporary professions. The professional body of esoteric knowledge yields a set of principles that become the guiding rules of good (that is, competent) practice for the conduct of the professional. Knowledge and practice converge in professional conduct, but they are not reducible to each other because the act of judgment necessarily intervenes. There is a certain cool deliberateness and remoteness about Kantian moral theory and the prescription for the relationship between knowledge and action. In his formulation, knowledge application becomes embodied in the idea of ethically and rationally disciplined conduct.

The goal of an applied "positive science" that would transform society was probably most unambiguously formulated in Auguste Comte's ideal of the social engineering. Ultimately, sociologists would be the high priests of the sociological religion—a conception that, like Kant's, also finds contemporary, albeit more sophisticated, manifestations and representatives. Since Comte was quite certain that positive science uncovered the truth, it was absolutely necessary that it guide society by transforming the mentality of society and man. Just as Comte had an uncomplicated conception of the nature of science, his unbroken faith in the power of science and positive reason led him to view the problem of knowledge application as essentially simple as well. Positive religion would promulgate a scientific mentality for the positive society in which scientific truth, once known, would direct action. It appears that Comte accepted certain limitations on the purposes of science that followed from this conception. Science was to serve social harmony; it must not overspecialize. While the Comtian program reflects an unambiguous belief in, and commitment to progress, it also emphasizes social statics and is directed more to the harmonious perfection of existing social arrangements than the scientific exploration of the unknown in even more specialized ways. There is a certain diffuseness about the Comtian program, in spite of its disarmingly naive simplicity, as indicated by the tension between the need for specific social engineering and the need for general positive religion. The problem of applying knowledge to human beings as objects and the problem of transforming mentality through conversion on the basis of knowledge both inhere in the Comtian conception.

The revolutionary program for the unity of theory and practice emerging from the Marxian tradition is quite different indeed. The philosophers of the Enlightenment firmly believed that a just social order could be established through human reason as long as discussion was free and once education was reorganized on the basis of the "science of ideas." To these thinkers, prejudice had its sources in vested interest, while the correct understanding of society resulting from enlightened knowledge application was a straightforward consequence of unprejudiced inquiry. This easy optimism was totally abandoned by Marx.

Marx criticized devastatingly the useless theorizing and spinning of abstract ideas on the part of people who had no practical involvement in society; those isolated specialists of ideas, "the professors." His purpose was to create scientific knowledge of the real nature of society. He believed that the unity of theory and practice existed when thought presses toward realization; reality itself must press towards the thought. Reality with respect to society seems to mean the transformation of consciousness that occurs in powerful mobilizations for sociopolitical action in the Marxian conception. From Marx's perspective, social theory is itself a component of the historical process; ultimately, the test of a theory is revolutionary practice and the advancement of the movement of history. The Marxian conception of the applicability of social knowledge emphasizes its capacity to transform consciousness, to mobilize political forces within the context of specified objective conditions. Knowledge and practice merge in the revolutionary act.

Both the Comtian and Marxian conceptions, in spite of their enormous differences, share a peculiar totalism. In both of these analyses, the social conditioning of ideas is perceived as a source of error and bias and fragmented consciousness, whereas valid knowledge is seen as essentially unitary in nature.

A very different understanding of these matters is found in the work of Max Weber. Weber emphasized the importance of ideas for an understanding of social life and considered ideas as an irreducible factor in society; yet in contrast to both Comte and Marx, he did not have a strong faith in the totalist conceptions of all history. But Weber's hiatus between fact and value, between knowledge and action or practice, between the scholar and the politician, is not a reflection of the indifference of "the professors" that Marx condemned. Rather, it is evidence of Weber's simultaneous commitment to scholarship and action and his lifelong effort to achieve and chart a creative relationship between these domains.

Weber conceived of modern societies as necessarily pluralistic in their values. Science produces knowledge within the structured and at least relatively autonomous domain of scientific inquiry and theory, but the needs and objectives of action involve commitments to other values. The problem of knowledge application in this conception, then, is essentially one of value clarification and the calculation of means.

While science and action, especially political action, necessarily belong to different realms, they are still related. It is the role of science to contribute to man's control over the environment by calculating both human activities and external objects; specifically by presenting precise calculations of the likely outcomes of an action. It further contributes to planning by indicating alternative modes of approach and courses of action, permitting a calculation of costs both in material and ethical terms. Most importantly, by displaying the characteristics of an action situation in relation to the alternatives being considered, scientific knowledge can play possibly even a decisive role in value clarification. Finally, science can contribute different methods and disciplines of thinking to the analyses of decision makers.

Weber emphasized, probably in an exaggerated manner, the irreconcilability of ultimate value positions. Yet he argued that science, including social science, finds its application in mapping the horizons of action and showing the actual implications that would causally follow from specific value informed decisions. And like Comte, Weber saw certain implications for science that resulted from the desire to use it in an ethical manner. However, his conclusions were drastically different: in order for science to be usable, it must retain and defend its autonomy and insist on its unique method and discipline. Therefore, he argued for "value free science"—a slogan that has frequently been misunderstood as a demand for ivory tower isolation in science when in fact Weber's argument for scientific objectivity was motivated by the desire to guarantee a critical role for applied science in value clarification.

A more recent perspective on the sources and uses of knowledge derives from the social construction of reality formulation represented by Peter Berger. His analysis of psychoanalysis and industrialized social structure attempted to assess systematically the relationship between the enormous popularity of psychoanalytic thought in America and American institutional practices. Berger's general premise was that social theories are built into institutional structures or arrangements and come to codetermine the social realities they explain. He suggested that this is especially important in the case of certain psychological theories, such as psychoanalysis, which define—in the sense of construct—the psychological realities they explain and interpret by providing a vocabulary and imagery for their expression.

> Since human beings are apparently destined not only to experience but also to explain themselves, we may assume that every society provides a psychological model (in some cases possibly more than one) precisely for this purpose of self-explanation. Such a psychological model may take any number of forms, from highly differentiated intellectual constructions to primitive myths. And once more we have here a dialectical relationship. The psychological reality produces the psychological model

insofar as the latter is an empirical description of the former. But the psychological reality is in turn produced by the psychological model, because the latter not only describes but defines the former, in that creative sense of definition intended in W. I. Thomas' famous statement that a situation defined as real will be real in its consequences, that is, will become reality as subjectively experienced by the members of that society.

(Berger, 1965:33–34)

Thus, Berger's analysis emphasizes the fact that certain items of knowledge, especially social and psychological theories such as psychoanalysis, come to codefine social states of affairs because they become integrated into institutional structures and subjective experiences, which in other respects such conceptual items are supposed to describe and explain. In Berger's understanding, then, social theories become part of the processes that construct social reality and structure consciousness. And from this perspective, psychological models may operate in society as self-fulfilling prophecies in that they generate or construct the subjectively experienced realities they pretend to describe and interpret objectively.

In quite a different vein, we all are aware that scientific and technological knowledge become integrated into institutional designs and structural arrangements. Certainly, different principles and styles of both personal conduct and organizational design emerge around the technology of steam engines, fossil fuel, electric generators, atomic reactors, and electronic computers. For example, the organizational and institutional logic of combustion rests on centralization; thus, the use of combustion associated with industrialization in the West spawned the factory as the critical site for economic productivity and generated the unprecedented urbanization that took place during the nineteenth and early twentieth centuries. In contrast, the organizational and institutional logic of electricity as a source of usable energy involves decentralization; electrical energy can be transported across large distances without serious dissipation, since it need not rely on mechanical connections between source and utilization. In short, engineering designs, technological knowledge, and the scientific principles underlying them become integral parts of organizations and institutional structures in the sense that such knowledge and the factual constraints it represents, while initially specialized and esoteric, comes to be the taken-for-granted working knowledge around which interpersonal relations, institutional patterns, authority structures, communications patterns, and other social arrangements crystallize.

This interpenetration between principles of personal conduct, interpersonal relations, and organizational patterns, on the one hand, and developments in socially available bodies of knowledge and their consequences, on the other, follows from knowledge implementation. It does seem at least intuitively clear, moreover, that technological knowledge implementation differs qualitatively, and especially in its consequences, from the diffusion of social and psychological knowledge that provides the symbols and vocabulary in terms of which actor-subjects describe and explain themselves. Yet, there also are important similarities between the implementation of technical knowledge around which demographic trends, organizational designs, occupational structures, and social solidarities may form, and the kinds of social knowledge that express, describe, and explain the outlook of consciousness of such solidarities and their members.

SOCIETAL LEARNING AND THE QUEST FOR KNOWLEDGE APPLICATION

The problem of knowledge application in highly differentiated, advanced society arises, as we saw, in already specialized contexts. People who feel pressing needs for specific knowledge or expertise are often themselves experts in certain other domains. For example, the school superintendent who wishes to assess the efficacy of a new curriculum, the police chief who wishes to introduce a more effective and rapid communication system, the urban planner who needs to know about patterns of air and noise pollution, and the like are experts in their own right. In these particular circumstances, knowledge needs arise in highly specialized contexts. The problem of knowledge application then involves the differentiation of the society and its capacity for knowledge transfer.

However, there is a larger question to be raised. As societies develop highly elaborated knowledge systems with their complex knowledge storage and retrieval function, one may speak of the possibility of societal learning in an explicit manner. Of course, societies since human evolution began have learned from experience. But such experience was incorporated in traditions, rituals, implicit symbolic codes, and became largely usable through individual memories molded by social structure. Increasingly, the social apparatus of preparing records and accounts of history and formal depositories of knowledge brings more explicit, self-conscious and deliberate forming of policy based on such knowledge within grasp. At least it is a reasonable aspiration to think that societal systems for knowledge might be capable of such learning.

Daniel Bell has given this idea an interesting expression.

> ...if one adopts a Parsonian view of the different functional subsystems of the society, one can say that the intellectual system, as the guardian of traditional values, has been responsible for one of the main integrative functions of the society. But in the future, as a source of innovation and as a bearer of change, it begins to replace "the economy" in carrying out the "adaptive" functions of the society. The question is whether the intellectual system is capable of carrying such a double burden of being both integrative and adaptive.
>
> *(Bell, 1968:160)*

It is, of course, not certain that such a vision of explicit self-conscious social self-direction involving planned change will ever truly come to pass. However, it is not unreasonable to think of the emergence of the differentiated knowledge system as at least raising the issue.

CONCLUSION

This concludes our overview of different conceptions of knowledge application. It has shown how these ideas are interrelated with fundamental images of society, indeed conceptions of man. Given the centrality of the issue of knowledge application in these concerns, it is strange that it did not receive much explicit treatment in the sociological tradition that appears to be most relevant to it, namely, the sociology of knowledge. Since we plan to use conceptual tools and perspectives related to that sociological tradition, we now turn to some historical notes on it.

3

Some Historical Notes on the Sociology of Knowledge

This chapter lays certain foundations for what is to follow by presenting our overview of the sociology of knowledge as an intellectual current. Our relation to that tradition is complex. In some regard we differentiate what we have to say and contrast it with the classical sociology of knowledge; in other regards we draw on that tradition and build on it. How this is done will become apparent if the reader keeps in mind that this chapter presents an historical background for the systematic treatment of theoretical issues that follow subsequently. For that reason, our treatment of the history of this field of inquiry will be selective and brief. Our object is to provide background for what is to follow, not to write a treatise on the history of the sociology of knowledge.

THE INTELLECTUAL AND HISTORICAL CONTEXT

The intellectual perspective and problems that gave rise to the sociology of knowledge reflect a fundamental change in the nature of intellectual and social self-consciousness that took place in the West, particularly in Germany and France, during the nineteenth and early twentieth centuries. The totality of this transformation is too vast to be described in this book; it would require a major study of intellectual history in its own right. It laid the foundation for reflective social science and history, clarified the role of methodology in the human sciences, and raised the problems of historicism and relativism as central ones in the development of social theory. This major intellectual reorganization signals the rise of critical reflectivity about knowledge and society—and, indeed, about consciousness itself. Available interpretations of this metamorphosis, such as Hughes' *Consciousness and Society* (1958), have dealt with only partial aspects of the change. Even Remmling's (1967 and 1973) efforts at tracing the origins of the sociological "thought style" of the sociology of knowledge, important though they are, remain selective and limited.

The origins of this transformation in social theory and intellectual self-consciousness stem from the expansion of critical thought and skepticism. It was the rigorous search for truth that also led to the relativization of truth. The context was the debate with eighteenth century thought, with the legacy of the Enlightenment. Enlightenment thought upheld reason as the critical measure of social institutions and their suitability for Man's nature. It presumed that essential rationality could lead him to freedom. It looked toward the perfectibility of humanity and society. Being infinitely perfectible, people could secure for themselves greater degrees of freedom through social critique and institutional modification. This in turn would enable them to actualize their potentially creative powers. But these powers were inhibited and repressed by existing institutions; being tradition bound and nonrational, they were not in accord with Man's basic nature. Enlightenment thought was therefore critical as well as scientific. The premises of rationality and the perfectibility of humanity and society eventually inspired the revolutionaries of France. After the revolution many of the most influential thinkers in Europe attributed the causes of that great upheaval to the Enlightenment *philosophes* and their ideas. Some intellectuals attempted to repudiate them. This response often has been treated as the era of Romanticism and Conservative Reaction. In general, the postrevolutionary period constituted an interesting and critically important phase in the development of social theory; it was out of this historical context that the sociology of knowledge in the formal sense emerged.

The legacy of Enlightenment thought was the common sense of progress. As things improved over time, the optimism about rationality and the applicability of knowledge to social institutions became deeply ingrained. Descartes, exponent of faith in rationality, had thought it possible to transform political and economic institutions in accord with reason. Voltaire presented a general history of civilization, conceived of in naturalistic terms. Montesquieu believed himself to have established that processes of politics in social life are subject to general laws, much as those that are found in other sciences. This framework of beliefs was shared by the encyclopedists who concluded that anything that interfered with freedom of thought must be opposed in the interest of humanity. Immanuel Kant spelled out his conception of Enlightenment with a similarly strong emphasis on freedom of thought and the independent judgment of the enlightened individual. There was an impetus to unmask prevailing prejudices based on religious or political interests. It was believed that scientific inquiry into the principles of society and morality would overcome such prejudice and reach the truth. But driving inquiry further turned it inward on itself; the enormous diversity of history and cultures and the complexity of human emotional life became sources of skepticism about the very universality of specific standards of reason.

One incisive intellectual event was the rise of critical inquiry in philosophy and, later, that of dialectical reasoning. Kant's dramatic question, "How is nature possible?" had focused attention on the structure of the human mind, on the human observer through whom nature is "synthesized." The insistence with which Kant made the structure of human consciousness and rationality itself the focus of his investigation was certainly not the only impetus to the self-consciousness of nineteenth century thought, but it was an important one. We will see that the Kantian problem and theme arise repeatedly, albeit in transformed ways, in the modern sociology of knowledge. Kant believed that the human mind is structured in a universal pattern of organization somewhat akin to the universal structure of the human body. Knowledge was therefore necessarily universal in nature, even though all knowledge was distinctly and specifically human.

The argument that knowledge and rationality evolve in the historical process from simple to complex forms, yet maintain some pattern of unity throughout this progression, was the outcome of a number of developments: Hegel's dialectic edifice of concepts; the rise of historiography, the discovery of systematic records of the past; the emergence of methodologies for investigating patterns of past knowledge; and, the construction of explicit methodologies and understanding of their relation to the structure of knowledge. The idea of a dialectic development of human reason through history relativizes the knowledge any given human knower possesses to a specific place in the encompassing context of the vast historical movement that Hegel saw as directed toward the consummation of the absolute spirit. Thus, the cool Enlightenment rationality of Kant, with its emphasis on the active synthesis of rational knowledge through the categories of the human mind, came to be superseded by dialectical reasoning dynamically oriented toward a conception of human progress to reason. Hegel's achievement was to formulate a conception of diverse views existing in the context of an overarching and unfolding unity of ideas and reality called, in the language of the times, "the absolute spirit."

A conception that linked fundamental issues concerning the certainty of knowledge not merely to either rational first givens or direct empirical experiences, but to the evolution of human mentality in the historical process remains a profound challenge. It forced social theorists to undertake a methodical search for firm foundations of knowledge, especially social knowledge. The problem was cast not only in terms of the historical bases of rationality, but also in terms of freedom, necessity, and constraint. The basic issues remain vital for contemporary theory, as Bershady (1973:30–31) noted in his study of Talcott Parsons' contribution. Following Parsons' analysis, he emphasizes that only part of Hegel's system was vigorously attacked, namely the idea that there is a continuity in the dialectical principle providing a structure of historical change. However, the underlying idea that historical data should be organized around the conception of a *Geist* and a unique historical constellation or system that is related to it, remained in place. As Bershady pointed out, the consequence of this was that such works stressed the wholeness, the "organic totality" characteristic of a system of meanings, and thus gave prominence to its uniqueness and its individual character. This mode of thinking thus led to the repudiation of general theory and even of general analytical concepts. This was so because "any attempt to break down this phenomenon [*Geist*] into elements that can be assumed under general categories of any sort destroys this individuality and leads not to valid knowledge but to a caricature of reality" (Parsons, 1937:480).

The emphasis on this kind of understanding influenced the main nineteenth century German intellectual tradition to stress an empathic immersion into historical cultural constellations focusing on their idiosyncratic uniqueness. This historicist position, grounded on painstaking historiographic craftsmanship, posed the radical challenge concerning the relativism of knowledge. Many significant nineteenth and early twentieth century sociologists felt compelled to respond. Increasing scholarly awareness of newly uncovered historical materials and interpretations of the intellectual productions of prior ages and civilizations also contributed to a growing critical awareness of the social relatedness of knowledge and exacerbated the emerging problem of ideology.

RELATIVISM AND THE PROBLEM OF IDEOLOGY

European intellectuals became concerned with the sociocultural and historical contexts that shape and color cultural productions such as knowledge. This awakening generated a growth of suspicion that the manifest surface and content of historical phenomena may be misleading; that what was presented as historical and political knowledge might be deception—intentional or unwitting—and that new kinds of understandings needed to be reached. Remmling (1967: 44f) described the emergence of critical concerns with knowledge and its social and historical contexts as the "road to suspicion," the title of his book on modern mentality and the sociology of knowledge. He argued:

> The spread of such intellectual and moral relativism [Remmling's reference is to historicism] accelerated the growth of the suspicion of ideology which began to invade the serene world of classical idealism when Francis Bacon declared that human reason was not clear light and set out to develop a fairly comprehensive list of disturbing influence with his enumeration of the

idols of the mind. Less indirect but not less significant for the
development of ideology was the reversal of classical idealism
that occurred when the modern idealist philosophers placed
"ideal reality" in the *consciousness of* man. . . . With Karl Marx
the suspicion of ideology becomes *total:* now the *entire* mind is
seen to be ideological—not just certain ideas; now the social situation does not merely condition the psychological manifestation of concepts but penetrates to the *noological* meanings, to
the very core of thought products. Marx turns this total suspicion
of ideology against the thought of his adversaries; the suspicion
of ideology becomes the most important weapon of the proletariat in the class struggle.

Marx struggled to establish a scientific understanding of
history as an unshakeable foundation not only of his theoretical
knowledge but also of historical action. To him the notion that ideas
as such govern the history of mankind and direct its evolution appeared an aberration—a professorial construction. History was shaped
by real forces, not by books. These real forces resided in the experience of men at work, concretely validated by the constraints and
necessities of daily experience. In the occupation world of work and
in its economic power constraints, he saw the core of social reality. It
was in this sense that he considered life as the determinant or foundation of consciousness, not the other way around.

Marx's great discovery was the foundation of idea-systems in
the solid ground of social structure. He saw them as depending on
arrangements for control of work and production through ownership; these he viewed as the most important constraints of concrete
human experience.

Marx's entire methodology emphasized the importance of the
material conditions of human life and experience. He criticized (in
the theses on Feuerbach) the older materialistic doctrines that understood reality only in the form of objects of contemplation. Against
that he emphasized the importance of "sensuous human activity," or
practice. For Marx, the question of what is truth was more a practical question than a theoretical one. Men are not only changed by
their circumstances; they also change them. Therefore, "the coinci-
dence of the changing of circumstances and of human activity or self-
changing can only be comprehended and rationally understood as

revolutionary practice" (Marx, 1968:145ff). Again and again Marx emphasized that "all social life is essentially practical. All the mysteries which urge theory into mysticism find their rational solution in human practice and in the comprehension of this practice" (ibid.). The fault of the philosophers was that they merely contemplated and interpreted the world: "The point is to change it."

Marx wrote the most succinct statement summarizing his sociology of knowledge:

> The general conclusion at which I arrived and which, once reached, continued to serve as the leading thread in my studies may be briefly summed up as follows: In the social production which men carry on they enter into definite relations that are indispensable and independent of their will; these relations of production correspond to a definite stage of development of the material powers of production. The sum total of these relations of production constitutes the economic structure of society—the real foundation, on which rise legal and political super-structures and to which correspond definite forms of social consciousness. The mode of production in material life determines the general character of the social, political, and spiritual processes of life. It is not the consciousness of men that determines their existence, but on the contrary, their social existence determines their consciousness. At a certain stage of their development the material forces of production in society come into conflict with the existing relations of production, or—what is but a legal expression for the same thing—with the property relations within which they had been at work before. From forms of development of the forces of production these relations turn into their fetters. Then comes the period of social revolution. With the change of the economic foundation the entire immense superstructure is more or less rapidly transformed. In considering such transformations the dinstinction should always be made between the material transformation of the economic conditions of production, which can be determined with the precision of natural science, and the legal, political, religious, aesthetic, or philosophic—in short, ideological—forms in which men become conscious of this conflict and fight it out. Just as our opinion of an individual is not based on what he thinks of himself, so we cannot judge such a period of transformation by its own consciousness; on the contrary, this consciousness must rather be explained from the con-

tradictions of material life, from the existing conflict between the social forces of production and the relations of production.

(Marx, 1904:11f)

The immense superstructure of symbolic, and for Marx necessarily "ideological," forms is an outgrowth of human labor. Men bring the products of their labor into relations with each other and transform them into values. However, they do not see that such things are simply the "receptacle" of human labor . . . "whenever, by an exchange, we equate as values our different products, by that very act, we also equate, as human labor, the different kinds of labor expended upon them. We are not aware of this, nevertheless we do it." (Marx, 1906:85) The human character of labor is perceived in society as an objective character of the products themselves. The reification of human creations that constrain human activity is difficult for men to penetrate. This remains so in spite of what Marx has believed to be his great scientific discovery of the underlying process. In analogy, the scientific discovery of the component gases of air did not change the way we experience the atmosphere. For Marx, the central dynamic of historical change was class conflict structured around the distribution of interests. This resulted from the institution of ownership of the means of production. The result was a theoretical construction capable of stark simplification. The subtlety of the Marxian conception of the relations between consciousness and social systems combined with a fundamentally simple vision of the historical process posed a compelling challenge. Certainly the problem of ideology confronted by European social thought resulted significantly from this challenge.

There were, of course, other sources of reflection. Nietzsche unmasked the workings of resentment, the struggle for power, and other forces in Man's inner world. The discoveries of Freudian psychoanalysis raised questions concerning the psychodynamic sources of bias in what is taken as truth. The problem of ideology was thereby given an additional dimension beyond the Marxian proposals concerning the social structural foundations of thought. Rationality may, after all, be a disguise for something else. It may really be rationalization of ideology. In this fashion, the Freudian and the Marxian critiques of what is taken for knowledge have mingled in the Western intellectual tradition to produce a stance of general suspicion and skepticism. What is presented as knowledge may not be truthful, but rather a torturous, if unwitting, deception. Only the domain of natural science, because of its exactitude and precision, was exempted. Here, then, is the problem of ideology with which classical sociologists of knowledge in the German tradition had to deal. Its central focus and concern, it can be seen in retrospect, were epistemological rather than sociological. It was a sometimes passionate search for ways in which one might distinguish ideology from truth.

CLASSICAL SOCIOLOGY OF KNOWLEDGE

The Mannheimian discussion of idea systems carried forward a basically Marxian perspective toward ideas as manifestations of the interests of specific groups located at particular points in the social structure. Judicial, political, philosophical, artistic, and social theories derived their meanings from a group's experience. Other groups often do not and possibly even cannot comprehend this meaning without great effort and further experience. Mannheim, however, differed significantly from Marx in applying ideological analysis to all groups, including Marxists and the proletariat. He also suggested that all ideologies have at least a partial hold on the truth. Thus, Mannheim reduced Marxian thought to the same status as that of other idea systems. It was his intellectual rigor and impartiality that brought him to that ethical and epistemological relativism concerning his own sociological formulations, which one may refer to as *Mannheim's Paradox:* if all social knowledge is historically and socially relative, how can Mannheim's theory itself make truth claims? For Mannheim, then, the sociology of knowledge seeks the meaning of doctrines and belief systems in the life situations of the persons who express them. In short, ideas are examined as they function in the lives of authors and audiences who are involved in their construction, transmission, and acceptance.

Mannheim made a basic distinction between two kinds of idea systems: ideologies and utopias. For Mannheim, ideologies are intellectualized justifications for the existing social order espoused by the established. They are moral and quasi-scientific defenses, descriptions, and explanations that support the status quo. In contrast, utopian ideas transcend the existing order and represent a critique and an attack on existing society by socially progressive, rising forces. Utopian movements hold out a practicable and appealing future at the same time that they call into question the legitimacy of the existing system. Mannheim argued, moreover, that utopias provoke ideologies: they make it necessary for the existing order to provide an intellectual defense of its values and of its operation. In fact, Mannheim observed that one prevalent ideology is precisely that of the pejorative view of Utopia: it is criticized as a visionary but unattainable state of affairs. Finally, Mannheim acknowledged that utopias can, and often do, become ideologies. That is, in its status as a criticism of an existing order, the utopian system of ideas or imagery contains the bases for a new social order—which will ultimately transform the utopian ideas into its own self-serving, justificatory ideology against new utopian critiques.

This phase of Mannheim's work represented a direct continuation of certain features in the Marxian tradition: in consequence, he continued to see the need for the gross categories of social class analysis as well as a social relational foundation for criteria of cognitive validity. However, his hopeful suggestion that a free floating intelligentsia not caught in the web of massive vested interests would have an historically and critically adequate perspective no longer seems convincing—if it ever was. His political sociology of ideology and utopia linked these belief systems to the historical interests of different social classes, juxtaposing those of backward ideological orientation (the capitalists) with those of progressive utopian belief-systems (the proletariat). This approach has had some utility for political analyses; what this formulation adds to an understanding of the social foundations of knowledge itself remains in question.

Mannheim did not overcome relativism in spite of his new term *relationism*. Perhaps this is because he did not deal with the problem of frames of reference systematically and radically enough. For example, he maintained that "in the thought 'two plus two equals four' there is no indication as to who did the thinking, and where." By contrast, he feels that "certain qualitative features of an object encountered in the living process of history are accessible only to minds of a certain structure" (Mannheim, 1952:194). Indeed, the analysis of a problem of addition requires just as specific an orientation and symbol system as the analysis of a cultural object such as the rules of evidence in eighteenth century English law. All knowledge is relational; but given a certain perspective it becomes rationally reproducible. Indeed, the empirical fact of the translatability of frames of reference—subject to certain limitations—seems to indicate an encompassing evolution of rationality.

In a subsequent work, Mannheim himself turned to issues beyond the sociology of knowledge. In *Man and Society in an Age of Reconstruction* (1940), he formulated a comprehensive conception of a sociology of planning. Planning involves not only engineering style instrumentalities, but also transformations of man, of personality. His objective was to provide, on the grounding of sociology and with the help of psychological knowledge, "planned guidance" to the lives of people. While this and his subsequent works on the diagnosis of our time (Mannheim, 1943) and freedom, power, and democratic planning (Mannheim, 1950) strike many important themes for applied sociology, they are hardly a sociology of knowledge applications.

One may well be skeptical also, from today's perspective, about the other great design for a sociology of knowledge—Max Scheler's. His sociology aimed to overcome relativism by providing the foundation for a philosophical anthropology, which meant the philosophical synthesis and interpretation of all knowledge about man. Scheler postulated the absoluteness of certain values; in practice this turned out to be a less than compelling effort.

Philosophically, his sociology was based on an epistemology that claimed to be broader than that underlying empirical science. He found such a basis in the phenomenological approach formulated by Husserl. Phenomenological anlaysis, he thought, not only yielded a systematic understanding of all the different roads to knowledge open to man, but also an understanding of unvarying essences. Although much of Scheler's work was in any reasonable sense of the term sociological, his ultimate intent was metaphysical. Empirical science needed to be seen as one avenue to knowledge. In fact, Scheler conceived of science as in need of guidance by an image of man in order to be a servant to wisdom.

Karl Mannheim criticized this program by pointing out:

Phenomenology holds that it is possible to grasp supra-temporary valid truths in *"essential intuition" (Wesensschau)*. In actual fact, however, we observe considerable divergencies among the intuitions achieved by different members of the school. These divergencies can be explained by the fact that intuitions of essence are always dependent on the historical background of the subject. Most impressive among phenomenological analyses are those based upon traditional Catholic values—our civilization, after all, is very largely a product of this tradition. It must be stressed, insofar as Scheler is concerned, that he has already dissociated himself from a number of Catholic tenets. This, however, is less important in the present context than the fact that he is still profoundly attached to the formal type of thinking exhibited by Catholicism.

(See Mannheim in Remmling, 1973:188.)

This quotation is, of course, an example of the extrinsic critique of ideas so characteristic of the classical sociology of knowledge. Nevertheless, Mannheim makes an important point.

Substantively, Scheler asserted that an adequate sociological understanding needed to come to terms with the fundamental dualism between *Geist* ("mind") and *Leben* ("life"). Mind is the name for the realm of meanings and rationalities that have their own inherent and essential structure. Life, on the other hand, he saw as being determined by the motivating forces of basic drives: for instance, procreation, hunger, power—these he called the real forces. The way in which these forces work themselves out in society is examined by *Realsoziologie*. Thus, there is a large difference between cultural sociology and realistic sociology. The sociology of knowledge is an aspect of cultural sociology. A general theory of mind becomes a necessary basis for all cultural sociology, just as the general understanding of the way in which human drives propel action is the basis for *Realsoziologie*. There is a fundamental difference between the two realms in that the realm of meanings can only be tackled using the analytical skills of cultural analysis based on a theory of the mind. The purpose of sociology, however, is to understand and explain the interactions between these two realms. Mind in itself determines the essence of all possible cultural formations; it cannot determine their actual realization. In order to become effective in society, mind must be connected with some real drive or force. Mind can only influence the direction of social processes under certain social circumstances; it is not in itself a realizing force.

Scheler's sociology of knowledge went far beyond the problem of ideology. In searching for a valid foundation of social self-understanding, he attempted to overcome the limitations of both Marxism and positivism. Both, he felt, were fallacious. Scheler argued that it must be recognized that there are different types of knowledge and that each has its own criterion of truth; each exists in its own social context. He based his sociology of knowledge on the following axioms: the knowledge a person has of membership in society is apriori, not empirical, because consciousness arises in and through society. Further, the empirical information that one has of others is gathered by different processes, depending on the structure of the group. These processes constitute a scale of ideal types; they range from emotional identification, on the one side, to analogical inference based on rational reflection and reconstruction, on the other. Finally, there is an invariant law of order determining the relationships between the different types or spheres of knowledge and truth.

These domains and their proper sequence are grounded in the sphere of the holy, which is followed by the sphere of interpersonal cognition, the sphere of external and internal experience relative to one's own body and its environment, the sphere of life, and finally the domain of physical bodies. In this sequential description of ideal types of different spheres of truth, Scheler emphasized the grounding of meaning in the domain of religious experience.

In order to simplify sociological analysis, Scheler distinguished three main types of intentionally structured knowledge: "saving knowledge," "wisdom," and "knowledge for control." Thus, the intentionality of knowledge became Scheler's structuring principle. Saving knowledge is created and assembled for the purpose of salvation. *Bildungswissen* (or "wisdom") aims to perfect humanity and its style of life, and knowledge for control is knowledge and information assembled for the purpose of power.

Even though both Mannheim's and Scheler's contributions are major landmarks of scholarship, they remain disturbingly fragmented and tenuously related to contemporary issues. Both wished to overcome the problem of relativism; both, while using very different perspectives, dealt with the metaphysics of society and history.

THE DURKHEIMIANS

The French tradition of Comte and later Durkheim is the other major source of the classical sociology of knowledge. It proceeded to formulate a perspective rather different from that adopted in Germany. To Comte, it was evident that changes in social structure are intimately associated with changes in modes of knowing. His evolutionary theory was, in ways parallel to the Hegelian conception, informed by commitment to the idea of progress in knowledge. The positive age of industrial technology, which Comte perceived as the culmination of modernity, was an age in which scientific truth would reign. Comte had a direct and simple faith in the efficacy of science, particularly scientific sociology, to provide knowledge not only for technological advance, but also for the governing of society. Scientist-experts and technically oriented entrepreneurs were the pivotal groups in his image of modern society. Their specialized, factual knowledge coupled with the positive religion would provide the direction for social life.

In its bouyant optimism the Comtian conception is an antithesis to the Marxian image of capitalist society. While Marx's scientific socialism led through revolution to the perfection of society, Comte conceived of an evolutionary process resulting in establishment of the industrial division of labor and emergence of a scientifically self-conscious and guided society. Comtian positivism is technocratic. Science is the vehicle of progress, but once positive science is correctly grasped, the perfection of society is clearly in view and possible. The Comtian spirit viewed society as an organic whole in an evolutionary process; the proper stance of the student of society was to treat it as a set of external facts, just as other scientists treated their subject matter objectively. In a sense, Comte's analyses of the theological and metaphysical age are in the mode of sociology of knowledge analysis; that is, he described the social contexts of modes of understanding. His analysis of industrial society, however, became a prescription for a science of society that would be both an understanding and a source of moral obligation. There is an important difference between these conceptions of the relation between society and knowledge and those created east of the Rhine. Comte's faith in reason was uncomplicated. His conception of science, informed by the standards of natural science, was simple and clear. Consequently, he failed to break through to an analysis of the structure of consciousness.

Emile Durkheim began his career of sociological inquiry influenced by these views. Sociology was to be a science that deals with social facts and facts are things that are external and constraining to society and the observer as well. Subjectivity has no room, at first, in this conception. Moreover, a close relationship between descriptive sociological inquiry and the construction of a system of ethics consistently permeated and informed his work. The image of modern society depicted in Durkheim's treatise on the division of labor was one of integration through diversity and functional interdependence. Although acutely aware of the social pathologies of industrial society, he was inclined to celebrate the division of labor as an achievement of enormous intricacy that created the conditions for freedom through the establishment of "organic solidarity."

Roscoe Hinkle has asserted that Durkheim's major image of social change was that of an evolutionist.

> Generally and pervasively Durkheim was a social evolutionist: His basic substantive concepts (e.g., the *sui generis* notion of the social and society, social morphology, the internal milieu, collective representations, the socially normal and pathological, the typology of societal species, moral rules and institutions, anomie), his ideas of social causes and social functions, his defense and use of the comparative method, and his conception of the fundamental and unifying problem for sociological inquiry are all basically and inextricably dependent on and interwoven with a broad theory of social evolutionary change. Similarly, his specific inquiries into the division of labor, family and kinship, suicide, religion, punishment, occupations, the state, contract, education, morality, and knowledge are cast in terms of social evolution.
>
> *(Hinkle, 1976)*

His empirical sociological analysis forced Durkheim to ask questions about the ways in which social life is meaningful to its participants. In contrast to the German intellectual tradition, he did not start out with questions of meaning, but his investigations led to these issues nevertheless. The collective consciousness or collective conscience—the French phrase has an instructive dual meaning—exercises constraining controls over the individual. Durkheim argued that individual mentality is a reflection of collective modes of thought in collective representations. Durkheim continued to probe problems of social pathology in his study of suicide. He found himself compelled to delve further into the construction of social meanings, the bestowal of meaning on individual's lives and actions through social participation, and the destruction of meanings by such states as anomie.

Durkheim's search for a sociological answer to the problem of the source of meanings led him to the study of religion and, ultimately, to the foundations of his approach to the sociology of knowledge. There are parallelisms here with what we have noted about Scheler. Durkheim, too, concluded that religion was the central phenomenon for an unravelling of the domain of meanings. Religion, he was convinced, was a symbolic transformation of something real.

This, of course, is the main point of departure in his *The Elementary Forms of the Religious Life* (Durkheim, 1915). Much to Durkheim's surprise, some of the main points he was trying to make in this fundamental study were missed or misunderstood by his critics. In 1914, therefore, he published a brief statement of what he considered his main arguments, which we will summarize here. Durkheim begins by emphasizing the interrelatedness of individuality and sociality "for society can exist only if it penetrates the consciousness of individuals and fashions it in 'its image and resemblance.'" Much of our mentality, thus, is social in origin. Indeed it is the social whole that produces or shapes its parts. "Consequently, it is impossible to attempt to explain the whole without explaining the part—without explaining, at least, the part as a result of the whole." This whole is a product of collective activity and, indeed, the highest such product is what is called civilization. So, the understanding of the sources and courses of civilization is an inquiry into the specifically human; it is also a main objective of sociology. Indeed, the conception of civilization and the analysis of civilizations play a special role for Durkheim.

The theme of what one might call Durkheim's cultural sociology, the understanding of civilization of which the sociology of knowledge is a part, is then posed by focusing on what Durkheim calls the constitutional duality of human nature. This constitutional duality is the distinction, which he believes to have been made everywhere, between body and soul, between the temporal and the transcendent. "The body is an integral part of the material universe, as it is made known to us by sensory experience: the abode of the soul is elsewhere and the soul tends ceaselessly to return to it. This abode is the world of the sacred." Given his belief in the universality of this duality, Durkheim concludes that it cannot be pure illusion. There must be something real in this human condition that gives rise to this feeling.

He finds this reality in the fact that we, as sensory, material beings, are "necessarily egoistic." By contrast, the realm of ideas, thought, and morals is necessarily universalized. By their very nature, concepts and morality transcend personal ends. Sensations are individual, but concepts always are common to a collectivity of some kind. They rely on language, and language always is the result of a "collective elaboration."

These two aspects of man's life are opposed to each other, indeed contradictory. "There is in us a being which represents everything in relation to itself and from its own point of view: in everything that it does, this being has no other object but itself. There is another being in us, however, which knows things *sub specie aeternitatis*, as if it were participating in some thought other than its own, and which, in its acts, tends to accomplish ends that surpass its own." The tension and antagonism between these two aspects of consciousness Durkheim relates to Immanuel Kant's moral philosophy, which juxtaposes the law of duty or categorical imperative to our empirical sensitivity that must be, as Durkheim has it, "humiliated" by it.

Quite in the Kantian spirit, he continues by asserting that this duality also exists in the sphere of our knowledge. "We understand only when we think in concepts. But sensory reality is not made to enter the framework of our concept spontaneously and by itself. It resists, and, in order to make it conform, we have to do some violence to it, we have to submit it to all sorts of laborious operations that alter it so that the mind can assimilate it." Our understanding in the realm of concepts always transforms direct and immediate experience, abstracts from it, in terms of a universalizing scheme.

Still following the Kantian approach to the problem of knowledge, Durkheim examines and rejects two proposed solutions that have been extant in the history of philosophy. He calls them the doctrines of empirical monism and idealistic monism. Empiricism is rejected because "it has never been able to explain how the inferior can become the superior; or how individual sensation which is obscure and confused can become the clear and distinct impersonal concept; or how self-interests can be transformed into disinterest." Idealism fares no better at his hands. Idealism cannot account for sensation. Both of these theories try to bypass or even eliminate the problem without solving it. The Kantian solution, which affirms the existence of two faculties, one of sensitivity and one of reason, he also finds wanting—even though he follows the Kantian line of argument quite far. It is insufficient, Durkheim argues, to give purely verbal answers; one must not be satisfied with postulating man's mental nature as some sort of ultimate given. On the contrary, it needs to be accounted for.

It is precisely this accounting for man's mental nature, civilization, and thus also for knowledge that he attempted in *The Elementary Forms of the Religious Life.* According to this analysis, "sacred things are simply collective ideals that have tied themselves on material objects. The ideas and sentiments that are elaborated by a collectivity, whatever it may be, are invested by reason of origin with an ascendancy and an authority that cause the particular individuals who think them and believe in them to represent them in the form of moral forces that dominate and sustain them." As a consequence of this origin, we feel reverence, respect, and even awe towards them. For sociological analysis, however, such "ideals are not due to any mysterious action of an external agency; they are simply the effects of that singularly creative and fertile psychic operation—which is scientifically analyzable—by which a plurality of individual consciousnesses enter into communion and are fused into a common consciousness." It is in this process that symbols are created that are external manifestations, carriers of meaning. The objects that have assumed symbolic significance embody collective representations and thus "arouse the same feelings as do the mental states that they represent and, in a manner of speaking, materialize." Here is one reason that religious symbolism is of central importance. Such ideals, which are products of group life, penetrate the individual consciousness and structure it. "Once the group has dissolved and the social communion has done its work, the individuals carry away within themselves these great religious, moral and intellectual conceptions that societies draw from their very hearts during their periods of greatest creativity."

In concluding this essay, Durkheim returns to the theme of the duality of human nature, emphasizing that he has shown how this duality corresponds to actual reality.

Durkheim's remarkable essay summarizes his perspective and also offers a rationale of a rather fundamental kind for his interest in the social origin of human understandings and symbolic forms. For example, he explored with Marcel Mauss the problem of the origins of classification. He poses the problem thus:

> . . . to classify is not only to form groups; it means arranging these groups according to particular relations. We imagine them as coordinated, or subordinate one to the other, we say that some (the species) are included in others (the genre), that the former are subsumed under the latter. There are some which are dominant, others which are dominated, still others which are independent of each other. Every classification implies a hierarchical order for which neither the tangible world nor our mind gives us the model. We therefore have reason to ask where it was found. The very terms which we use in order to characterize it allow us to presume that all these logical notions have an extralogical origin. We say that species of the same genre are connected by relations of kinship; they call certain classes "families"; did not the very word *genos* (genre) itself orginally designate a group of relatives (γενοσ)? These facts lead us to the conjecture that the scheme of classification is not the spontaneous product of abstract understanding, but results from a process into which all sorts of foreign elements enter.
>
> *(Durkheim and Mauss, 1963:8)*

From this vantage point, Durkheim could feel that he had superseded Kant. After all, Kant proclaimed that the human observer sees the world through the organization of human understanding and reason—through innate and unvarying categories of the mind. Whereas Kant believed that he had derived these categories from his study of the accumulated lessons to be learned from the history of logic, Durkheim felt that Kant had merely proclaimed them. In contrast, the Durkheimian hypothesis about the relation between social structure and symbolic structure provided a scientific key for the comprehension of the categories of the mind. To Durkheim and his coworkers, this hypothesis represented a scientific and philosophical breakthrough of the highest order.

Although the Durkheimian conception of the relation between symbolism and social structure has proved fruitful, it is not the only key to understanding meanings in their social context. The early Durkheimian hopes for a simple derivation of symbol systems from social structures were disappointed—even though major correspondences between, for example, religious belief systems and the distribution of authority in a society were subsequently demonstrated by Swanson. The Durkheimian tradition of searching for structures of symbolism has been continued in the structural anthropology of Levi-Strauss that seeks, similar to the thrust of structural linguistics, deep structures in the cultural symbolism of societies.

Our brief sketch of the two somewhat independent strands of development leading to the early sociology of knowledge shows important convergences. The concern with understanding meanings, with their explication and explanation, has been a pervasive one in the history of sociology and social psychology. Indeed, it is compelling to see how serious social theorists with the divergent perspectives of Marx, Mannheim, Scheler, and Durkheim have in some fashion agreed on the centrality of meaning for social analysis. (Obviously there was no consensus among them on procedure; methodology for the sociology of knowledge did not keep pace with theoretical development.) In the early period, analytical techniques had to be fashioned by theorists individually as they proceeded to deal with the questions that substantively intrigued them. Progress on analytical issues is relatively recent.

MAX WEBER

One of the most important problems for such methodological progress was relating conceptions of specific individual actions, experiences, and meanings to large-scale sociocultural phenomena. This difficulty forced many analysts into logically questionable short cuts in forming concepts and propositions. Max Weber treated subjectively meaningful conduct (action) as the basic unit of sociological analysis. In his view, social relations and institutional orders can be described as probabilities for the occurrence of specific kinds of meaningful actions; the construction of meanings occurs in action. This approach opened up the methodological passage for analytical conceptions and procedures that would link the analysis of meanings with the study of large-scale social structures. At the same time, Weber did not concern himself with the detailed empirical investigation of the building blocks for this analysis, namely, the social psychology of individual action itself. The concept action remained for him a theoretically basic, if unexplicated, construct.

This was because Weber built his conceptual apparatus for a theory of action not to create a closed system of concepts, but to assemble the tools he believed necessary for investigation of the major questions that principally concerned him: the reasons for the development of modern capitalist society and the direction of that vast historical movement in the West that he called the process of "rationalization." Although he exercised great skill in the task of analyzing the life worlds of the actors he studied, trying to discover their essential attributes and structures of meaning, he did not develop a general theory of symbolism or a systematic sociology of knowledge.

The study of the life forms of the great civilizations, the dynamics of the world religions, was the major project among Weber's sociological investigations through which he pursued the understanding of the sources and fate of modern Western capitalist civilization. The quest for an understanding of the encompassing process of rationalization in Western civilization led Weber into a broad ranging comparative enterprise. In the author's introduction to his collected essays, on the sociology of religion, Weber succinctly expressed these considerations (Weber, 1958b:13ff). "A product of modern European civilization, studying any problem of universal history, is bound to ask himself to what combination of circumstances the fact should be attributed that in Western civilization, and in Western civilization only, cultural phenomena have appeared which (as we like to think) lie in a line of development having *universal* significance and value."

Rational and organized science appeared only in the West—even though empirical knowledge, philosophical reflection, and theological wisdom certainly are not a monopoly of the West. A similar phenomenon occurred in the world of art.

The musical ear of other peoples has probably been even more sensitively developed than our own, certainly not less so. Polyphonic music of various kinds has been widely distributed over the earth. The co-operation of a number of instruments and also the singing of parts have existed elsewhere. All our rational tone intervals have been known and calculated. But rational harmonious music, both counterpoint and harmony, formation of the tone material on the basis of three triads with the harmonic third; our chromatics and enharmonics, not interpreted in terms of space, but, since the Renaissance, of harmony; our orchestra, with its string quartet as a nucleus, and the organization of ensembles of wind instruments; our bass accompaniment; our system of notation, which has made possible the composition and production of modern musical works, and thus their very survival; our sonatas, symphonies, operas; and finally, as means to all these our fundamental instruments, the organ, piano, violin, etc.; all these things are known only in the Occident, although programme music, tone poetry, alteration of tones and chromatics, have existed in various musical traditions as means of expression.

(Ibid.)

Similar phenomena of rationalization can be found in other aspects of culture in the West, for example, in architecture. The trained official and bureaucracy are a specifically western development.

And the same is true of the most fateful force in our modern life, capitalism. The impulse to acquisition, pursuit of gain, of money, of the greatest possible amount of money, has in itself had nothing to do with capitalism. . . . One may say that it has been common to all sorts and conditions of men at all times and in all countries of the earth, wherever the objective possibility of it is or has been given. It should be taught in the kindergarten of cultural history that this naive idea of capitalism must be given up once and for all. . . . Capitalism *may* even be identical with the restraint, or at least a rational tempering, of this irrational impulse. But capitalism is identical with the pursuit of profit, and forever *renewed* profit, by means of continuous, rational, capitalistic enterprise.

(Weber, 1958b:17)

Rationalization has many facets and aspects. There is the rationalization of the economy, the rationalization of administration and law, even, as we have seen, the rationalization of music. It is a vast process in the transformation of meanings, not only in the direction of developing greater instrumental skills, but in the sense of emerging rationalized life-styles. Indeed, Weber's enterprise took him into the very center of the sociological understanding of meaning and thus also knowledge.

With regard to the central importance of religion, he was of one mind with Durkheim. Weber also gave great weight to the study of social inequalities in the form of stratification systems and the distribution of power. Unlike Durkheim, however, Weber did not suspect that religious meanings were simply direct reflections of social structure. He had a much more complex appreciation of the interdependence between structural constraints faced by individual actors in social positions and their vital interests, values, and conceptions of ultimate reality.

Weber viewed the process of rationalization as an enormous differentiating process of social evolution that could not be reduced to a simple formula of either class conflict or collective representations.

He retained an intense sense of history and appreciation for the way the idiosyncratic and unique effect subsequent history. He was suspicious of simplifying, generalizing formulas. But he did describe one aspect of rationalization as a process of increasing differentiation among specialized groups, coupled with increases in the symbolic capacity of specialists within these groups.

In the sociology of religion, for example, we are made aware of the significance of religious virtuosos and experts as well as of the differentiation and increasing sophistication inherent in an organized, scholarly theology. In the sociology of law, we are made aware of processes involved in building a profession that becomes increasingly separated from the surrounding society and begins to define its own technical domain of competence, until it establishes itself as a semi-autonomous, major force in modern societies. In the sociology of science, there is a similar emphasis on structural differentiation and increasing specialization of the frame of reference adopted by scientists. Similar themes can also be found in Weber's analysis of the impact of accounting and the codification of technical skills in book-keeping on the economy and on the construction of enterprises separated from households. Again, the emergent theme is the differentiation of rationality into specialized domains in the treatment of bureaucracy.

In brief, Weber treated social differentiation as an aspect of evolution that is linked to the production of rational forms of conduct in specialized ways by technical working groups such as lawyers, accountants, bureaucrats, and scientists. All these groups carve out autonomous domains of meaning and proceed to create a language, systems of control, and standards for work and conduct appropriate to themselves. However, the core of the rationalization process lies in the transformation of basic meaning structures, and indeed forms of life. It is certainly not exhausted by the emergence of specialized forms of rationality, even though these are an important phenomenon especially in the contemporary world.

Weber's conception of modern society is complex and therefore difficult for the casual student. Rationalization also means the progressive "disenchantment" of the world. Towards the end of the *Protestant Ethic and the Spirit of Capitalism* (Weber, 1958b:181f), he writes:

> Since asceticism undertook to remodel the world and to work out its ideals in the world, material goods have gained an increasing and finally an inexorable power over the lives of men as at no previous period in history. To-day the spirit of religious asceticism—whether finally, who knows?—has escaped from the cage. But victorious capitalism, since it rests on mechanical foundations, needs its support no longer. The rosy blush of its laughing heir, the enlightenment, seems also to be irretrievably fading, and the idea of duty in one's calling prowls about, in our lives like the ghost of dead religious beliefs. Where the fulfillment of the calling cannot directly be related to the highest spiritual and cultural values, or when, on the other hand, it need not be felt simply as economic compulsion, the individual generally abandons the attempt to justify it at all. In the field of its highest development, in the United States, the pursuit of wealth, stripped

of its religious and ethical meaning, tends to become associated with purely mundane passions, which often actually give it the character of sport.

No one knows who will live in this cage in the future, or whether at the end of this tremendous development entirely new prophets will arise, or there will be a great rebirth of old ideas and ideals, or, if neither, mechanized petrification, embellished with a sort of convulsive self-importance. For of the last stage of this cultural development, it might well be truly said: "specialists without spirit, sensualists without heart; this nullity imagines that it has attained a level of civilization never before achieved."

GEORG SIMMEL

Weber did not give the explicit treatment of knowledge the same centrality that Georg Simmel assigned it. Simmel clearly grasped the peculiar nature of society as an object in a way that escaped many other early students of sociology. For him, the nature of social processes, association, was in large part cognitive. Knowledge of social circumstances and knowledge itself are constitutive of society. Persons in interaction act as social observers *and* agents and it is the way they know—or think they know—others and social entities that is an important determinant of their relationships and social structures. In his famous essay on the presuppositions of sociology, Simmel asked the question, "How is society possible?" In the critical style of Kant, he wished to examine the conditions under which society can be known as against those under which nature is known. Again, the significance of the Kantian mode of questioning crops up for our enterprise.

Kant had demonstrated that nature was "synthesized" by the observer and constituted through such a synthesis in terms of the categories of the mind. Society, Simmel argued, already contains its own synthesis. Its units, subdivisions, and structural arrangements are in part determined by the conceptions that its members have of themselves, of others, and of their social relationships. Knowledge of social affairs is constitutive of social structure, but not in the sense that all social activity can be reduced to its cognitive aspect; there are often motives and desires in social circumstances beyond those known to the actors. But knowledge is a key determinant of social structure. It is interesting that, in Simmel's phrasing, the question asks not merely how social structure determines knowledge but rather how modes of social knowledge become determinants for social structure.

Simmel insisted that the study of the patterns and forms of social cognition is a subject of critical sociological importance. Society consists of conscious persons. It already contains social knowledge before the scientific observer begins to construct a personal knowledge of it—and indeed this observer must perforce rely in large measure on the social knowledge actors are willing to reveal. Therefore, the science of sociology cannot proceed in quite the same way as experimental physics. It must deal with the complexity and the special requirements imposed by the inescapable circumstance that sociologists study knowledge-containing systems within which persons orient themselves to others in terms of discernible perspectives. Simmel drew two major conclusions from this insight. The first was that the sociologist needed to clarify and articulate his/her own perspective so that it would be clearly understandable as a specialized systematic way of mapping social reality. The second was that it is important to study social cognition not only in its structure within an individual, but also in the resulting structured distributions of social knowledge. It is for these reasons that Simmel wrote about subjects like "knowledge, truth and falsehood in human relations," "types of social relationships by degrees of typical knowledge of their participants," "secrecy," and other aspects of the creation, distribution, and systematic withholding or falsification of social knowledge.

His classic investigations of lying reveals fundamental aspects of social relationships. Simmel correctly perceived that lying is considered a graver crime in modern societies than in traditionally structured ones. Trust and truthfulness are the social cement without which the integration of highly differentiated societies is endangered; even claims to the validity of authority depend on its ability to meet demands for truthfulness.

These remarks do not imply that Simmel thought of modern society as governed by unshakeable commitments to truth; rather, he identified the particularly central importance of cognitive trust in modern societies, which necessarily contain a variety of cognitive perspectives that become differentiated along with an increasingly complex division of labor. The contemporary concern with mistrust and deception is a reflection of the centrality of cognitive trust in modern life. Seen from this perspective, Simmel's investigation of secret societies—their structure and forms of control over members and their influence on society generally—is not merely a romantic excursion into exotic social phenomena. The problem of secrecy and its significance naturally comes into focus for a sociologist who gives primacy to the cognitive domain.

One of the most important studies on social cognition in the sociology of knowledge is Simmel's massive *The Philosophy of Money*. Money is examined as a medium of symbolic measurement and exchange, the introduction of which produces major alterations in modes of social cognition and thus consciousness. In this work, Simmel unfolded an explicit concern with the understanding of symbolism as it facilitated and constrained social cognition and communication and shaped emerging structures of consciousness. The Simmelian view of society, then, is that of a network of structured interactions among conscious and reflective actors who form views and images of themselves as they form images of others. These actors participate in a flow of communication in which they construct "syntheses," typifications of the attributes and boundaries of roles, groups, and societies. Without being bound by Simmel's specific interests and conceptual framework, we follow his imagery in this work.

THE SOCIAL CONSTRUCTION OF REALITY

We have attempted to describe, in painfully abbreviated and simplistic fashion, certain perspectives of the classical sociology of knowledge and of great sociologists who contributed to the emergence of the sociology of knowledge and exerted a dominant influence on this field for a long time. In the postwar period, a different perspective began to emerge, a perspective that has been called "the reality constructionist" frame of reference. This perspective acknowledges the legacy of the classical sociologist, especially the

Marxian tradition, but it should emphasize a special indebtedness to other theorists such as Georg Simmel. Its most important representative was Alfred Schutz. The central issue of the analysis of meanings, especially as posed in Max Weber's sociology, was taken up in the work of Alfred Schutz. He sensed correctly, we believe, that the understanding of meaningful action as it was sketched by Weber and even Simmel was incomplete; that a more specific and detailed theory of action was needed. Schutz set for himself the task of developing such a highly specific, empirical basis for understanding action and its relations to meaning. He drew both on the work of American pragmatists such as William James and George Herbert Mead and on the work of European phenomenologists such as Husserl in attempting a close-up study of action. Of particular relevance for our formulation is Schutz's contribution to the clarification of the relation between symbol, reality, and society.

In his classic essay "On Multiple Realities," Schutz proceeded from William James' famous analysis of our sense of reality. To James, the origin of reality was subjective; the "real" is what excites and stimulates our interest in a certain way. But James uncovered several "sub-universes" of reality and proceeded to sketch their structure. They include the worlds of sense for physical things, of science, of ideal relations, of the idols of the tribe, of mythology and religion, of individual opinion, and of sheer madness.

From this recognition of the existence of multiple realities or finite provinces of meaning, Schutz proceeded to describe the everyday world of working as the paramount reality. This reality of daily life has a particular structure that is maintained on the basis of sometimes precarious arrangements. The philosopher, and particularly the phenomenologist, practice skepticism and may indeed suspend belief in apparent reality. Man within the "natural attitude" of the common-sense reality of everyday life suspends all doubt that the world and its objects might be other than they appear to him.

There are many domains of reality that Schutz called "finite provinces of meaning." There are the worlds of dreams, of fantasy, of art, of religious experience, of scientific knowledge, and others. Each of these finite provinces of meaning has a peculiar cognitive style and each has a particular mode of reality characteristics to it. Transitions from the paramount reality of everyday life to other finite provinces of meaning may be experienced through reality shocks. This Schutz-ian approach to the study of action, symbols, and meaning became an important intellectual stimulus to sociology. The challenge of understanding reality as socially constructed has been taken up as a major foundation of the contemporary reality constructionist perspective in the sociology of knowledge.

Among these developments, the emergence of studies and theories of behavior that emphasize the cognitive domain were particularly relevant. The early work of Von Uexküll exemplified the revolution in behavioral biology brought about through the systematic study of the way in which different species perceived their environment and behaved in relation to it. In psychology, the Gestalt tradition supplemented and gradually merged with the behavioral learning tradition, becoming the contemporary cognitive psychology endeavor to investigate the domain of meanings with quantitative precision. This line of investigation deals quite specifically with the domain that Scheler believed to be inaccessible to scientific observation. In addition, philosophical interest in the study of symbolism and language, represented by the massive work of Cassirer, brought technical analytical tools to the investigation of cultural objects, meaning systems, and modes of inquiry and discourse.

Probably most important is the fact that the earlier classical sociology of knowledge, certainly in the era of Comte and Marx, but even in the period of Mannheim and Scheler, lacked analytical techniques that have since been developed, ranging from multivariate statistical analysis to model logics and other formal tools for the explication of meaning systems, including the capability to model inquiry and cognitive processes on computers. The classical sociology of knowledge had to solve methodological problems by intuition and informed guesses as to where the solution to a problem or any inquiry might lie. No field can ever do without such intuition, but the proliferation of techniques for the study of symbolism, including the emergence of a science of linguistics, has created a much more promising intellectual climate.

In brief, intellectual interest in the role of various kinds of knowledge in social processes and in social structure, while still embedded in philosophical issues of great scope, can now be systematically pursued with the conceptual and procedural apparatus that social science has constructed. In this context, the focus has shifted from the preoccupation with ideologies and belief systems to the study of bodies of knowledge and their specific consequences for social structure. While the sociology of ideology, as the older version of the field should more properly be called, focused on the social sources of epistemological error—as in Mannheim's conception of ideology for example—the sociology of knowledge approach adopted here deals with cognition and knowledge as central aspects of the social process. It thus investigates both the social context in which knowledge is created and the consequences it has for social life.

PART

I

Bibliographic Notes

Bibliographies essentially have to do with proof and therefore reflect some underlying epistemological position. Our concern is not to demonstrate that we have produced truths about contemporary knowledge but to help generate and make a responsible contribution to the awareness of those interested, as well as make possible responsible reactions to our own interpretations. The footnotes and bibliographic notes that follow, then, are shaped by this sense of an emerging shared intellectual enterprise. The problem of producing a bibliography for a topic such as "the knowledge system in post-modern society" is oppressive considering that there are all manner of overlapping areas of discussion that have considerable relevance. A systematic bibliography, moreover, would require a book of its own, and that is not the task to which we have addressed ourselves. Instead, we used these bibliographical remarks to indicate the contributions we were most centrally influenced by, interested in, and responsive to, even though they may reflect epistemological, sociological, or substantive positions that diverge from our own.

Chapter 1 centers on the emergence of a new kind of society, which we refer to as post-modern, and suggests that this development is accompanied by significant changes in the cultural system of common sense. Among the numerous works that discuss this phenomenon, Daniel Bell's *The Coming of Post-Industrial Society* (1973), *The Cultural Contradictions of Capitalism* (1976), and his article "Teletext and Technology: New Networks of Knowledge and Information in Post-Industrial Society" in *Encounter Magazine* (Summer, 1977) represent a fundamental contribution. The same kind of theme is developed, from an entirely different point of view, by Alain Touraine in *The Post-Industrial Society, Tomorrow's Social History: Classes, Conflicts and Culture in the Programmed Society* (1972). A still different perspective can be found in Amitai Etzioni's *The Active Society: A Theory of Societal and Political Processes* (1968). A related, but generally distinct, body of writings has attempted to describe and interpret the more cultural dimensions of contemporary society. These include Berger, Berger, and Kellner's *The Homeless Mind: Modernization and Consciousness* (1973). Vytautas Kavolis's illuminating article "Post-Modern Man: Psycho-Cultural Responses to Social Trends" (1970), and a number of the essays included in the volume edited by Robertson and Holzner entitled *Identity and Authority* (1979). Two works dealing with the role of knowledge in contemporary society bear specific mention: one is Fritz Machlup's pioneering study, *The Production and Distribution of Knowledge in the United States* (1962), which addresses the subject from an economic frame of reference, and the other is Robert E. Lane's article on "The Decline of Politics and Ideology in a Knowledgeable Society" (1966). The treatment of post-modern common sense draws heavily on an unpublished manuscript by Clifford Geertz entitled "Common Sense as a Cultural System," which extends Geertz's earlier treatments of "Ideology as a Cultural System" and "Religion as a Cultural System" (1964 & 1966) to a new symbolic domain. Finally, statistical support for many of the arguments and interpretations advanced in Chapter 1 are drawn from the U.S. Department of Commerce, Bureau of the Census volume *Historical Statistics of the United States: Colonial Times to 1970* (1975), the Bicentennial Edition, Parts I and II.

Chapter 2, "The Quest for Knowledge Application," draws on an array of references dealing with the nature and types of knowledge in society. Among this array of materials, several loom especially large in our formulation: Burkart Holzner's *Reality Construction in Society* (1972), Peter L. Berger and Thomas Luckmann's *The Social Construction of Reality* (1967), and Berger's articles on "Psychoanalysis and the Sociology of Knowledge" and "Identity as a Problem for the Sociology of Knowledge" (1965 & 1966) constitute a strategic point of departure. A different point of departure into the same subject matter can be found in Robert K. Merton's collection *The Sociology of Science: Theoretical and Empirical Investigations* (1973), edited by Norman W. Storer. Some basic works reflecting the growing concern with knowledge application are Lazarsfeld and Reitz's *An Introduction to Applied Sociology* (1975), the collection of essays edited by Raymond E. Bauer entitled *Social Indicators* (1966), and the array of articles produced by Michael Radnor and his associates, such as the 1975 piece, "Analysis of Comparative Research, Development and Innovation Systems and Management: With Implications for Education."

Sections of this chapter deal with historical conceptions of knowledge application. Here we draw heavily on the 1959 translation of Comte's *The Positive Philosophy* by H. Martineau, the 1968 edition of the translation of Marx's *Theses on Feuerbach* by Rudolf Miller, the Shils and Finch translation of Max Weber's *The Methodology of the Social Sciences* (1949 & 1966), as well as the 1958a translation of Weber's essay "Science as a Vocation" that appears in the Gerth and Mills volume, *From Max Weber.*

Chapter 3 presents some historical notes on the sociology of knowledge. A large number of important contributions to this area appear in the volume edited by James E. Curtis and John W. Petras on *The Sociology of Knowledge* (1970). Among the many works that treat the intellectual and historical contexts within which the sociology of knowledge emerged, several are of particular significance: H. Stewart Hughes' *Consciousness and Society* (1958), Karl Löwith's *From Hegel to Nietzsche: The Revolution in Nineteenth Century Thought* (1964); and Gunter W. Remmling's *Road to Suspicion: A Study of Modern Mentality and the Sociology of Knowledge* (1967), as well as his 1973 collection of readings entitled *Towards the Sociology of Knowledge: Origin and Development of a Sociological Thought Style.* A recent contribution to discussions in this area is Harold J. Bershady's *Ideology and Social Knowledge* (1973). The major figures and view points that contributed to the classical sociology of knowledge have been amply interpreted in a number of works, of which Robert K. Merton's original essay in the collection *Social Theory and Social Structure* remains an important landmark. Two other landmarks in the emergence of this sociological specialty are Karl Mannheim's *Ideology and Utopia: An Introduction to the Sociology of Knowledge* (1936) and his *Essays on the Sociology of Knowledge*, edited by Paul Kecskemsti (1952).

A Conceptual Framework and Tools for a Sociology of Knowledge

Matters that we have considered in Part I—historical changes in cultural systems, traditional models of and for knowledge application, the emergence of the sociology of knowledge—will be relegated to the background in the two chapters that comprise Part II of the book. In these chapters, we present the conceptual framework within which we will proceed, as well as the specific theoretical tools we will employ in our subsequent analysis of the contemporary knowledge system and knowledge application. Thus, chapter 4 introduces our perspective on the social construction of reality as well as some critical theoretical constructs in some detail. We suggest that these constructs represent the fundamental bases for any serious, systematic attempt to develop a sociology of knowledge and cultural systems. Among the conceptual tools developed in this chapter, we devote particular attention to symbolism, frames of reference, and those specialized work groups that rely on a body of theoretical knowledge that we refer to as epistemic communities. Chapter 5 takes up the critical question of the relation between social structure and knowledge as a basis for our examination of the knowledge system in Part III. In this chapter, our interest is in the social structures as frameworks and repositories for knowledge. This leads us to consider the

*manner in which bodies of knowledge are differentially distrib-
uted throughout the social structure; we even suggest the fruitfulness of viewing social structure as the patterned distribution
of bodies of knowledge among various social positions. The subjects and concepts discussed in these two chapters, then, provide the necessary armamentarium for the analysis of the post-
modern knowledge system that constitutes the third part of the
book.*

4

The Perspective: The Social Construction of Reality

ON THE EXPERIENCE OF REALITY

Sociological concern with basic social processes underlying the construction of shared realities frequently is viewed with considerable alarm. The assertion that reality is socially constructed somehow threatens to deprive it of its very nature, its firmness, or its reliability. This sociological concern does indeed differ from the common sense conception of everyday reality, which accepts the world as given, as what men take it to be, and as a simple object of deliberate attempts to use, transform, or adjust to it. From the perspective of common sense, the world of everyday life is taken for granted as *reality*. It is simply, compellingly, and self-evidently *there*. To entertain doubt about its reality would require a suspension of one's everyday routines and meanings. A chaotic tumult of events and situations, lacking both interpretations and interpretability, would threaten to overwhelm the individual. Nevertheless, recent work on the socially constructed nature of the reality of everyday life—a foundation for the sociology of knowledge on which our presentation relies heavily—is actually a systematic extension and continuation of basic insights in the core sociological tradition. We especially build on the works of Alfred Schutz (1967), Berger and Luckmann (1966), and Holzner (1968 and 1972).

Among all of these Schutz is especially important. Indeed, the idea of the social construction of reality has been sharply formulated in his work. His important book *The Phenomenology of the Social World* (1967) is a translation from the German. In fact, the original German title might also have been rendered as "The Meaningful Construction of the Social World." Because apprehensions and misunderstandings are so abundant when these issues are raised, we will attempt to introduce the reality constructionist frame of reference with deliberate, step-by-step care.

Sociology since the time of Weber and Thomas has taken it as axiomatic that people act in terms of the subjective meanings that situations have for them. Following W. I. Thomas' observation, situations that are perceived and experienced as real are real in their consequences for what people do. Clearly the manner in which a Buddhist monk, a Soviet functionary, or an English scientist define the same situation will be markedly different. Thomas' dictum merely serves to orient us to the fact that these divergencies or discrepancies in situational definitions have consequences for subsequent social behaviors of those who experience them. Stated more simply, the meaning a person attaches to a situation determines how that person will act in the situation. That meaning is not given in the nature of the objective environment; it does not inhere in its characteristics. It is, rather, the result of active definition.

This observation is straightforward; we know that the life worlds of various people differ, that they have different modes of understanding and evaluating, and that therefore the structure of their experience, knowledge, and reactions will be characteristically different. Individuals assign a variety of meanings to their environment in order to interpret and render it meaningful for their actions. For the sociological observer, these meanings provide primary and inescapable sources of empirical data. There is no doubt that what a professional gambler, a Western scientist, and a Benedictine monk take for reality, in general, or how they interpret any particular situation is different in content, quality, and texture. Yet until recently, sociologists have hesitated to take the problem of reality construction, maintenance, and transformation seriously. They have attempted to dissolve it into issues of divergent attitudes, values, or personality psychodynamics, rather than address the problem of divergent experiences of reality.

Such reluctance is understandable. Every one of us, sociologists no less than others, experiences the reality of the situations we encounter as directly given and compelling. In this sense, reality inheres in the objects and situations themselves; it is, as Durkheim emphasized, both external and constraining. Similarly, we are certain of the concreteness of our own corporeal reality and extend this sense of firmness to many matters that we know about and whose existence it is unreasonable to doubt. This natural and unreflected, *given* conception of reality is also pervaded by a sense of its unitary structure. It is the reality of a unitary, coherent, external, and constraining world of objects (things) and situations (places) in which we live as unitary selves with an unmistakeable sense of subjecthood and personal identity. True, this world can be explored and we can make discoveries about it—but each discovery confirms that indubitable existence of that particular reality as well as the paramount fact that it preceded our existence and will continue to last beyond us.

This certitude of reality as unitarily, objectively there—and certainly not merely subjectively imagined or socially constructed— provides a firm framework for the manner in which we perceive any particular thing. It is extremely difficult to sum up this attitude by observing it across many diverse situations, and then reflect on it. Yet precisely this accomplishment is necessary for the practitioner of sociology who employs a *reality constructionist* perspective. Just as the unitary, coherent, and constraining reality of an existing world and the observer's subjectively experienced personal identity are indubitably given, so too is the fact that there are other observers, subjects, who also experience a whole and unitary world and a coherent integrated identity of their own. Systematic attention to the processes by which observer and subjects come to share the reality of their worlds and identities—producing the intersubjectivity of reality— is the fundamental basis for the study of social action from our frame of reference.

When a subject's reality is scrutinized critically, we discover that it does not actually possess the unitary, integrated character that endows its experiential aspect and permeates the subjectively experienced impression the subject has of that reality. Thus, if we question people about the circumstances under which, and the ways in which, they think of things as real, we discover enormous variability, so that the quality of experienced reality takes on a wide variety of shadings.

William James (1952: 641ff) raised these issues concerning the perception of reality when he observed that there are many subuniverses of reality, ranging from the world of the senses to that of the supernatural. Of course James was not discussing the ontological question of what reality really is—which he clearly put aside—but the empirical question of the experience of reality and unreality on the part of subjects. Nevertheless, he demonstrated that, on close and systematic scrutiny, an individual's account of his or her experienced reality referred to several differentiated domains—or discrete realities.

This differentiation of the experience of reality becomes even more apparent when we turn from the perspective of one individual to the distribution of reality experiences and experienced realities among members of a group or an entire society. The common sense observation that there are different points of view, ways of understanding, and characteristic interpretations between, for example, a child and an adult, or between an expert and a layman, clearly demonstrates that the differentiation of social life worlds by age, sex, or role status leads to a great differentiation in what is experienced as real. It was precisely this problem that Alfred Schutz addressed in his classic paper "On Multiple Realities" (1945:533–576). Schutz attempted to delineate "finite provinces of meaning," each having its characteristic cognitive style. In his analysis, the experiencing subject moves along, among, and between these finite provinces of meaning or subsectors of reality that have clearly circumscribed boundaries and modes of experience. These finite meaning sectors are, however, enveloped by the paramount reality of everyday life and it is to that reality that consciousness always returns after brief excursions to these more specialized and problematic sectors of reality.

This differentiation of what actors in social life take for real poses one of the most recalcitrant problems for any understanding that social observers may form of it. The sociological observer remains firmly grounded in the ontology of empirical science, taking as data intersubjectively verifiable observations recorded in terms of his conceptual scheme so that they become facts constituting his reality. However, it is the peculiar quality of the social domain observed that it contains conscious agents whose experienced realities differ from the observer's. That is, the manner in which conscious agents who are members of society perceive each other, the way in which they ascribe reality or unreality to different domains, itself is constitutive of the phenomenon the sociologist studies, namely society. The very fact that social structures can be shown to have a resilient facticity, that they are not easily amenable to change, rests in part on the fact that the persons living within them act in terms of hard and constraining realities they perceive and experience. For this reason, the statement that "reality is socially constructed" is not meant as an empty figure of speech; rather, it emphasizes the fact that those experiences to which actors attach the stigma *real* occur in socially allocated situations and are perceived and interpreted in terms of socially derived and validated as well as differentiated meanings.

The fact that we are focusing on reality experience and not merely beliefs, attitudes, or values is of critical importance. We need not become embroiled in the philosophical or semantic intricacies of the term reality to utilize it in sociological analyses. For our purposes, Berger and Luckmann's (1967:1) definition of reality as ". . . a quality appertaining to phenomena that we recognize as having a being independent of our own volition ('we cannot wish them away')" is quite adequate. And following this line of thought, we are arguing that "the sociology of knowledge must first of all concern itself with what people 'know' as 'reality' in their everyday, non- or pre-theoretical lives" (Berger and Luckmann, 1967:15). The social psychological study of the beliefs, attitudes, and values held by various categories of individuals is certainly an important issue, but it is a different issue from the sociological analysis of the shared construction and social diversity of experienced reality. Beliefs, attitudes, and values have been shown to be relatively amenable to modification through persuasion or alteration of the individual's point of view. In contrast, what people know as reality appears to be relatively difficult to change. The reason that subjectively experienced reality is so difficult to transform is that it rests on powerful plausibility structures and is constrained by the allocation of environments and social processes of *reality maintenance.* A most important reason for the stability of subjectively experienced reality is that it represents an internalized version of what is taken for granted as objective reality. That is, as a result of the emergence of selfhood in socialization, there arises a symmetrical relationship between objectively and subjectively experienced reality. Conversely, a near total transformation in a person's subjective reality, which Berger and Luckmann (1967:

156-163) refer to as "alternation," requires a comparable alteration in that individual's experience of objective reality, the way one experiences the shared realities in one's life. The gravity of this issue comes into focus when we deal with the problem of the social diversity of experienced reality.

It is now apparent that empirical sociology proceeds in terms of a special frame of reference whose ontology is, for the purposes at hand, not questioned but rather taken for granted. It is a frame of reference formed around the task of the systematic empirical description and explanation of purposive social action. However, because of the centrality of actors' cognitions and their reality experiences, the frame of reference underlying the sociology of knowledge requires attention to the manner in which such diversity and regularity in reality experiences on the part of members in society come about. Berger and Luckmann (1967:3) have indicated that the sociological interest in what is taken for reality and knowledge is basically justified by the fact of their social relationality and diversity:

> The need for a "sociology of knowledge" is thus already given with the observable differences between societies in terms of what is taken for granted as "knowledge" in them. Beyond this, however, a discipline calling itself by this name will have to concern itself with the general ways by which "realities" are taken as "known" in human societies.

They go on to argue that:

> The sociology of knowledge must concern itself with whatever passes for "knowledge" in a society, regardless of the ultimate validity or invalidity (by whatever criteria) of such "knowledge." And insofar as all human "knowledge" is developed, transmitted, and maintained in social situations, the sociology of knowledge must seek to understand the processes by which this is done in such a way that a taken-for-granted "reality" congeals for the man in the street. In other words, we contend that *the sociology of knowledge is concerned with the analysis of the social construction of reality.*

The problem with which we are dealing when we discuss the social construction of reality is not entirely unique to sociology. It is clear that the problem of meaning encountered by anthropologists in their study of the bases of cultural (especially religious) meaning systems is very closely related to the sociological problem of shared reality. Geertz suggests that it is the sense of analytic, emotional, or moral impotence—in the form of bafflement or ignorance, pain or suffering, and injustice or intractable ethical paradox—that constitutes radical challenges to the common sense view that life is comprehensible and that we can, by taking thought, orient ourselves effectively within it. He suggests that any cultural system must cope with at least these three challenges to meaning if it is to persist. Analogously, we are suggesting that any viable social system must cope with challenges to the maintenance of a common sense conception of the paramount reality of everyday, workaday life. This paramount reality embedded in common sense encompasses more specialized realities as finite provinces of meaning and action. In short, the cultural-anthropological interest in meaning systems meshes with, and is supplemented by, sociological considerations of the structural bases and institutional processes involved in the construction and maintenance of shared reality definitions, and both have consequences for social action (see Geertz, 1973:100–109). We will return to issues concerning divergent experience or definitions of reality, and problems of reality construction and maintenance, repeatedly in this and the following chapters because of their central importance in understanding knowledge application processes.

EXCURSUS: SLEEP AND THE PARAMOUNT REALITY OF EVERYDAY LIFE

It is in the wide-awake, everyday life that consciousness imposes itself in the most urgent and intense manner. As Berger and Luckmann (1967:21) note:

> I experience everyday life in the state of being wide awake. This wide awake state of existing in and apprehending to reality of everyday life is taken by me to be normal and self-evident, that is, it constitutes my natural attitude.

It is in this reality of everyday life that actors deal with the external world, adapt to it, or affect it. It is this relation of the reality of everyday life to effectiveness that warrants the designation of paramount reality. However, the boundary between what common sense defines as wide-awake reality and alternate states of experiencing needs to be delineated. We have already suggested that there are alternate modes of experience. It may help if we consider briefly the differences between the wide-awake, daytime reality of everyday life and the mode of experience in sleep and dreams.

Most adult people in most known societies sleep for about seven or eight hours every night in a place rigidly defined in relation to socially significant groups of others. During this time people enter a domain of experience in which the reality of wide-awake, everyday life is peculiarly suspended. There is significance in the fact that the term everyday life ignores the night, a significant portion of the twenty-four-hour cycle. The term is consistent with our cultural emphasis, which denies the validity, reality, and the responsibility of sleep life and dream experiences. We follow Aubert and White (1959: 46–54) in using the term "validity" to refer to the degree to which events in the sleep world are perceived as a true guide to the waking world of conscious choice and decision; the term "reality" to refer to the extent to which the sleep world is perceived as coextensive and similar to the waking world; and the term "responsibility" to indicate the extent to which individuals are rewarded or sanctioned for events in the nighttime world of sleep. Aubert and White (1959:49) conclude that "these three attributes apparently can vary independently, and in particular reality and responsibility do not necessarily go together." Much of our following discussion is based on our understanding of key works. We rely very heavily on Nathaniel Kleitman, *Sleep and Wakefulness* (1959). In addition, we have found the anthology by Ralf L. Woods (1947) very useful as well as the classical works by Freud (1932) and Fromm (1951).

There is striking variation between cultures with regard to these matters, but Western societies deny the validity, reality, and responsibility of dream experience and sleep life. The paramount, wide-awake, daytime reality of everyday life is considered the only socially significant one. The circumstance that the wide-awake, daytime reality is treated *as* reality—or as the only reality worth attending to—in societies within the circle of Western civilization is related to the importance of intersubjective reality tests resting on consensual validation and the norm of universalistic achievement. Specifically, the central legitimate justification for social action in these societies is made by reference to intersubjectively shared, testable performances or experiences achieved by an individual, rather than by references to inherent, ascribed personal qualities or cues and/or assistance from extra human forces. Since the domain of experiences one enters when one falls asleep at night is not shared with others, it is impossible to verify it intersubjectively through consensual validation and, since it is not a performance gearing into the external world, modern Western societies have tended to classify events that occur while asleep as invalid and unreal imaginings for which the actor cannot possibly be held responsible.

Yet if one attends to the mode of experiencing, it is undeniable that sleep constitutes a domain with altogether different dimensions that cannot be described using designs derived from the linear, homogeneous time of the intersubjectively shared, external world. From the point of view of the sleeper, a night may seem as long as the most infinite eternity or as brief as a single moment's unconsciousness. Moreover, waking time and sleeping time are perceived and experienced as unfolding differently. Waking time is conceived of within an essentially developed mental frame of reference. It is perceived and experienced as cumulative; the individual and those around him/her are seen as growing, developing, and aging in terms of time spent awake in the reality of everyday life. Time asleep is largely ignored; it seems excluded from development. In this sense it is treated as basically noncumulative time and perceived in static terms. As Aubert and White (1959:51) put it, in this culture sleep time is perceived:

> ... as a scene of life to which one returns "in the same place" over and over again. Comparing the two images of time, the irreversibly vanishing stream and the monotonously recurring pointer enclosed by the circle of the clock, the former image appears more associated with daytime, the latter image more with sleep-time—the wasted time. By this cultural definition, sleep-time becomes a redeemer from the fears associated with the inevitable passing of time and the definitive loss of the past, "le temps perdu." In more than one sense, sleep represents an encounter

with the past, a recurrent regression, whereas daytime impresses upon actors the necessity to decide on what to leave behind as parts of an incontrovertible past.

This definition of the reality of sleep life is far from universally shared in different cultures. Many cultures develop elaborate techniques in order to secure dreams of a prescribed content as a requirement for an initiation rite or as a means of divination connected with illness or witchcraft. Moreover, a candidate for the role of Shaman may work very hard in order to produce the proper dreams at specific, culturally designated times and locations. The apparent success of such efforts indicates that dreaming may well be a learned activity—in a sense that transcends the by now trivial observation that Freudians dream Freudian dreams containing Freudian symbolism, Jungians dream Jungian dreams complete with Jungian archtypical symbols, Adlerian dreams feature appropriate symbols, and so on.

There are other socially constructed and culturally specific characteristics that differentiate daytime reality from the experience of sleep life and dreams. For example, especially advanced modern cultures contain the idea that to awaken from sleep is to begin afresh, to start anew with past performances at least partially canceled. This is because the premise that sleep is a different kind of time means that each day's events are qualitatively bounded and, hence, have their own origin and significance. In this sense, they are not merely viewed as constituting a discrete section to be filled within an endlessly ongoing stream of events. In modern Western societies, this conception encourages at times more flexible, adaptive behavior and reduces one's sense of guilt and the need for other personality defenses. Again Aubert and White observe (1959:52):

> Without a loss of status, an actor may start on a new sequence of behavior, even if it violates the norms which he seemed to support on the previous day when interaction went into the latency phase. He "has slept on it" and started "a new day."

A related theme in Western culture concerns innovation and creativity. Nighttime nonresponsibility and flexibility with respect to perception or action are defined in this culture as contexts in which demands for conformity and protection of traditional status are lowered considerably. Hence, both nighttime activities and the experiences during sleep—namely dreams—have been associated with creative and innovative production.

One other aspect of the boundary between sleep and daytime experiences involves the fact that transitions to and from the physiological state of sleep are often perceived as hazardous. This is at least due in part to cultural definitions that surround the role of the sleeper. For example, there are fears of death or not awakening; fears of separation from the everyday world and loneliness; fears concerning helplessness and passivity; fears about unacceptable inner impulses that may break through into consciousness or into behavior that takes place in the wide-awake, daytime reality of everyday life. Last, as Berger and Luckmann (1967:21) put it:

> As I move from one reality to another, I experience the transition as a kind of shock. This shock is to be understood as caused by the shift in attentiveness that the transition entails. Waking up from a dream illustrates this shift most simply.

THE SOCIAL CONSTRUCTION OF REALITY

The manner in which reality is constructed, its domains separated from each other and bounded, and knowledge about it intersubjectively circulated and exchanged, is not a random or haphazard process. On the contrary, we suggest that major advances in understanding human affairs can be made by attempting to discover the lawfulness in these processes. Indeed, one might speak metaphorically of the "grammar" of reality construction; yet, we are not suggesting that it is entirely or even predominantly a linguistic process.

All domains of sharable experiences and reality are relational in the sense that shared orientations mesh with situations to produce interpretative frames within which reality becomes experienced and known. There is thus an inherent perspectivity in the process, which itself becomes a condition for gaining access to a domain of reality. Yet such perspectives are indeed accessible and, under certain circumstances, translatable into each other.

We will pursue these matters in relation to the concept of knowledge by exploring knowledge as a sociological construct and introducing further conceptual tools we consider essential. Therefore, we will deal with symbolism, with frames of reference, with reality tests, and with those special collective phenomena predominantly structured around modes of knowing, namely epistemic communities. An exploration of those domains in which the structures of authority, status, and compelling common sense are partially lifted, which we call *cultural free spaces*, will conclude this overview.

KNOWLEDGE AS A SOCIOLOGICAL CONSTRUCT

The objective of analyzing the social construction of reality is modest and empirical. It does not attempt to question or even to study ultimate ontological issues. Rather, the objective is to describe and analyze the lawful processes by which human actors determine what is real for them and how they form knowledge from such experiences. Although it is clearly necessary to distinguish the perspective of the observer (in this case the sociologist) from that of the people being studied, this is not a difficult matter in principle. The problem is to determine the social context in which knowledge is formed, maintained, changed, distributed, and utilized.

Some argue that the term *knowledge* should be used only to denote a veridical representation of a state of affairs. This requirement cannot, in principle or in practice, be accepted by the sociology of knowledge. It is true that the objective adequacy of knowledge, as it may be tested in terms of its effectiveness, for example, is a matter of decisive import for humanity's power over nature. However, it is not easily possible for the sociological observer to ascertain whether this is the case or not. Further, the observer's focus is on what, in a given sociocultural context, is taken for knowledge. While the sociologist accepts the frame of reference of empirical inquiry, he/she deals with a domain of phenomena in which the criteria for testing knowledge applied by the people studied vary considerably. Indeed, it is the sociologist's task to describe and explain the use of such criteria and their consequences. Thus, knowledge as the subject matter of sociological inquiry is simply what a group or a society takes as knowledge. This is not a matter of infinite variation. Certainly what is taken as knowledge by one group may be taken for superstition or error by another; but this, too, is not subject to random or unaccountable variation. In fact, we believe that discovering lawfulness in these relations will contribute to a more general theory of rationality, rather than strengthen a historicist withdrawal into relativism.

Within this sociological framework, knowledge can be defined as follows:

> . . . "knowledge" can only mean the "mapping" of experienced reality by some observer. It cannot mean the "grasping" of reality itself. In fact, philosophical progress has produced the conclusive insight that there can be no such thing as the direct and "true" apprehension of "reality" itself. More strictly speaking, we are compelled to define "knowledge" as the communicable mapping of some aspect of experienced reality by an observer in symbolic terms.
>
> *(Holzner, 1968:20)*

This definition emphasizes a certain inescapable relationality of all knowledge. All knowledge is that of an observer who communicates a symbolic mapping of some experienced reality so that it is accepted as factual. The reality referred to may be of an empirical, pragmatic, or often normative or supernatural order; the manner in which the observer gains access to this reality through experience may vary dramatically. Yet all knowledge originates in an observer and retains the stamp of that observer's peculiar relation to the experiential base. The fact that knowledge must be symbolically represented and communicated also means that it must be structured in terms of some particular symbol system that is, in turn, linked to the observer's stance and experience. Symbolic representation is constrained by the requirement for communication to some broadly or narrowly conceived audience that, in turn, serves to validate the knowledge presented.

Knowledge conceived of in these terms is clearly the core and the focus of reality construction processes. Our conception of knowledge includes that of science, but is certainly not limited to it. Furthermore, we recognize that such factual knowledge, whatever its domain, is conceived of by those who hold it as embedded in an aura of beliefs held with varying degrees of certainty. Addressing ourselves to those reality constructs taken as facts brings us closer to the hard core of social structure and social process than does focusing on the more ephemeral aspects of beliefs and their uncertainty.

SYMBOLISM

Man's social world is, above all, a world of symbols. Myth, religion, art, and science are cultural realities created in terms of symbolism. As Cassirer (1944:24) put it:

> Man has, as it were, discovered a new method of adapting himself to his environment. Between the receptor system and the effector system, which are to be found in all animal species, we find in man a third link which we may describe as the symbolic system. This new acquisition transforms the whole of human life. As compared with the other animals man lives not merely in a broader reality; he lives, so to speak, in a new *dimension* of reality.

Building on the stimulus of the new behavioral biology of Von Uexküll (to whom we referred earlier), Cassirer saw the possibility of a new philosophical theory of man founded on the analysis of symbolism and the symbolic transformation of experience.

The symbolic representation of meanings creates the possibility of stabilizing them. Without symbolization, an organism is restricted in its responses to the stimuli immediately given in its environment or to genetically fixed response patterns. The capacity to create and use symbols permits human beings to transcend the limitations of the immediately given environment, to represent objects in their absence, to imagine objects that have not yet existed or will never exist, and to work on relations among them. In other words, Cassirer pointed to symbolization as a human social characteristic that has enabled mankind to expand its cognitive domain, its experiences, and presentations of reality by an enormous measure.

Universal applicability, owing to the fact that everything has a name, is one of the greatest prerogatives of human symbolism. But it is not the only one. There is still another characteristic of symbols which accompanies and complements this one, and forms its necessary correlate. A symbol is not only universal but extremely variable. I can express the same meaning in various languages; and even within the limits of a single language a certain thought or idea, may be expressed in quite different terms.

(Cassirer, 1944:36)

The universal applicability and the variable uses that can be made of symbolism are among the factors that permit the stabilization of meanings. A conception of meaning, represented by a symbol, can be abstracted from the dizzyingly variable flow of concrete sensory experience. Human beings live in a reasonably stable world, at least to the degree that it is composed of definite objects with definite relations among them. Beyond that point, human sensory experience is like that of other animals, highly variable–affected by shadings of light, distances, and the internal state of the organism's moods or needs. A concrete perception is unique unto itself; that a perception can be integrated with previously learned meanings into a conception of an object is a result of the close integration of symbolic representations in the human cognitive process.

Symbolically defined objects are perceived as meaningful things. They are placed into some relatively stable framework that is symbolically representable and anchored. These objects are typically perceived as in juxtaposition to a self. The symbolization of objects implies a mode of symbolizing the experience of selfhood. Since objects and subjects are symbolically defined, human cognition is, in many subtle ways, regulated through the symbol system that adults have acquired. The symbolization of experience and object is linked, as George Herbert Mead pointed out, to the very structuring of the human mind. The use of symbolism in interaction enables a human to "take the role of the other," thus learning to view one as subject and object at the same time, albeit in different respects. Reflexivity, that is selfhood, appears to be built into the very structure of symbolism itself, as Langer implied (1942:49):

Symbols are not proxy for their objects, but are vehicles for the conception of objects. To conceive a thing or a situation is not the same thing as to "react toward it" overtly, or to be aware of its presence. In talking about things we have conceptions of them, not the things themselves; and it is the conceptions, not the things, that symbols directly "mean." Behavior toward conceptions is what words normally evoke; this is the typical process of thinking.

As vehicles for conceptions, symbols necessarily have some physical base, such as the sounds of the spoken word or the canvas and paint of a portrait. More importantly, they convey a cluster of meanings or conceptions that is necessarily abstract relative to any concrete sensory experience to which they may refer. Most significantly, symbols relate to each other in structured systems. The creation of new symbols and cognitive operations within a cultural system (for example, to investigate relations among concepts or meanings) follows a relatively stable structure. In the case of language, normative rules can be distilled by the grammarian and even more basic structural principles are discovered by the linguist.

The fact that symbol systems, such as languages, operate in terms of consistent structures defines realms of possible cognitive operations. Thus, the structure of a symbol system is both the enabling ground as well as a constraint on the construction of symbolic accounts for experiences. Such account rendering is not the only purpose of symbolism, but it is of special interest in relation to our focus on knowledge. For example, the vocabularies of languages develop in relation to situations and patterns of usage: vocabularies tend to reflect the structure of collective experience—differentiated though it may be—of a language community. The resultant language structure can become an important factor in facilitating or impeding social and cultural change. For example, Gallagher (1969:58ff) points out that modern Arabic has long tended strongly towards stabilization and away from accommodation to change through adaptation. By contrast, Bahasa Indonesian, a language that emerged only in the past generation from pre-modern and very simple structures, opened itself to innovation by borrowing vocabulary. He mentions that the difficulties of Arabic and other mid-Eastern languages in dealing with abstract political terms such as freedom, nationalism, state, nation, and such—were not experienced in the Japanese context. Gallagher goes on to investigate the differential receptivity of non-Western languages to the technical vocabulary and linguistic

structures of scientific discourse. In the recent past, Arabic functioned as a relatively closed, sacred language for the Muslim world;
hence, it developed great disadvantages with regard to adaptability
and receptivity. By contrast, Gallagher describes the cases of Bahasa
Indonesian and Japanese, of which the former is probably one of the
most prominent examples of a planned or artificial language. Gallagher concludes (1969:87):

> Ultimately the inter-connected problem of language and science
> reduces itself to the fact that it is impossible for a broadly scien-
> tific mentality to grow up in a society which has not become
> reasonably modern, that is to say, which has not at least gone
> some considerable distance along the road to modernism. It is
> impossible for a society to go very far along the road to modern-
> ism unless it possesses rationalized linguistic equipment that can
> be utilized in the journey.

Maybe Gallagher stresses the determining power of language
too much. Certainly we do not wish to assert that language forms
must be available preceding their use; however, language forms may
exercise a limiting, as well as a facilitating influence. The example of
the relationships discussed in Gallagher's work should be taken as an
illustration of the interdependencies between available linguistic symbolism and the forming of concepts—even though the relationship is
a less determinate one than Gallagher's formulation implies.

> The knowing observer must abide by certain rules of mapping
> which are essentially given by the structure of the symbolic sys-
> tem which he uses. The nature of the concepts, the nature of
> the permissible operations with these concepts, is in large mea-
> sure defined by the structure of the symbolic system itself, e.g.,
> by the language in terms of which he describes reality. There are
> several specific aspects of the structure of symbolic systems sig-
> nificant for "mapping." They are, for example, its internal organi-
> zation, the relation between symbols and the experiential refer-
> ents, and the requirements of communication.
>
> *(Holzner, 1968:21)*

The first point refers to the manner in which the internal
structure of the symbol system limits the forms of knowledge expres-
sible in that language; such limitations include the rules of logic or

the grammar of the language. In terms of such rules, consistency and contradiction are defined within the internal structure of the symbol system itself. The second point deals with the linkage between the symbol system and the reality it represents. All symbolism, we have seen, represents conceptions—but these in turn stand for experience of some kind. The question, then, is how a symbol can be linked back to the experience from which it was conceptually distilled. With symbols representing empirical knowledge, this is done through operational definitions that denote, presumably, the events originally observed. In the case of sacred symbolism, representing esoteric knowledge, it is necessary to reproduce the proper attitude of mind and prepare the observer for the privilege of access to the domain of experience for which the symbol stands. Many other possibilities exist, but the manner in which symbolic representations refer back to experience becomes an important constraint on their use. Finally, symbol systems are structured in terms of requirements for communication. In its most general form, this is quite obvious. Some observers even believe that symbolism has its primary function in communication—a belief we clearly do not share. Modes of communication and, indeed, their physical arrangements have a profound influence on the structure of symbolism used. This is obvious in the case of the communicative conventions of the scientific research paper or patterns of communication through the mass media. The very vehicle of communication provides occasions for the creation and redefinition of symbolism itself.

A final observation about symbolism remains to be made in this context. While we have emphasized those aspects of symbolism particularly relevant to the construction and communication of knowledge, symbolism obviously is of direct social significance. The two domains overlap, but there are specific dynamics to social symbolization, especially in the construction of individual and collective identities. Social arrangements are symbolically defined, for example, through rules of conduct, but they are also represented to constituent members and outsiders through what Emile Durkheim called collective representations and through landmark symbols that stand for collective entities. Surely the quality of such social symbolism through which, for example, the identity of a nation is presented is interdependent with the symbolic vehicles of social knowledge. There is a great difference between relatively unreflected symbolizations that signal, say, a social boundary around a group, and reflected symbolizations of social reality that stand for the results of social assessments and analyses. Through this linkage between symbolism representing knowledge and symbolism representing loyalties and identities, social knowledge acquires special significance.

FRAMES OF REFERENCE

Our conception of knowledge in its relationality to an observer—one's symbol systems providing means for mapping experience, as well as representing and communicating it—emphasizes that it is necessary to specify the observer's position and point of view. For this purpose we use the concept *frame of reference*. By frame of reference we mean a structure consisting of taken-for-granted assumptions, preferences for symbol systems, and analytical devices within which an observer's inquiry proceeds. This structure specifically defines the relation of the observer to what he/she knows and represents. For a rational understandability of knowledge, it is necessary to signal the frame of reference in use.

People in general interpret and analyze the objects and situations they meet in fairly systematic terms—it is a safe assumption that commonfolk as well as scholars tend to rely on what they consider rational or at least reasonable modes of inquiry and explanation. Any such inquiry is contextually situated. An observer in a social structure may sample only certain types of social occurrences. Others remain hidden from view simply because he/she is not in a position to notice them. Similarly, the physical observer needs to specify his/her location in time and space in relation to the objects viewed. Any specific inquiry as a deliberate process to construct and ascertain knowledge remains necessarily embedded in a context that itself is not the object of the inquiry. Thus, we speak of situated rationality both in social structural and ecological terms (which, so to speak, define the sampling frame of the observer and the allocation of situations to him/her) and in terms of frames of reference. The latter constitute the most proximate context of inquiry.

We use frames of reference to orient ourselves to specific objects, to conceptualize what problems are and how they are constituted, and to determine their possible or permissible solutions. This certainly includes conceptions of the manner in which facts are to be determined and what kinds of explanations may be constructed. Examples for such frames of reference from contemporary society abound in the world of the professions. There is the frame of reference of the trial lawyer who proceeds in the context of adversary proceedings before a court and jury, whose evidence is sworn testimony, and whose problem is defined by the legal system as the innocence or guilt of the client. There is the surgeon who may deal with the same facts as the researcher in cardiology—but in a different context and manner—or there is the social worker, whose frame of reference is often in conflict with that of an accountant.

Modern professions are, indeed, structured around frames of reference that define the specific professional mode of inquiry; they claim perspectives unique to the autonomous professional that are not accessible to clients or other outsiders. As a consequence, differences in frames of reference are often involved in professional-client tensions. Disputes about facts—as between sociologists and lawyers—can often be shown to involve divergent conceptions of the nature of facts. The sociologist may be highly suspicious of the use of sworn testimony as scientific evidence, knowing a great deal about the effects of group pressures, suggestions, and persuasion on the evidence witnesses give. The lawyer may be very suspicious of the results of a sociological survey, knowing a great deal about the unreliability and variability of hearsay evidence, no matter how large the quantities that are collected.

A general cognitive sociology must necessarily study frames of reference. This not only includes the frames of reference of professionals or specialists in the creation of symbolism and knowledge, but also the frames of reference in common sense use. This introduces a complexity into the application of the concept. Certain frames of reference, while always having the status of the context of inquiry and therefore not reflected on during its progress, are explicitly codified and articulated. Other frames of reference remain implicit and lack specific symbolic articulation. The modern professionals will tend to be able to discuss their frames of reference with much greater precision than traditional craftsmen can. The latter's frame of reference rarely becomes an object of inquiry. The former not only receives much attention, but also is dramatized to underscore the significance of the professional's unique perspective and point of view.

In the study of knowledge in post-modern society, one often encounters highly articulate and elaborate frames of reference. It is therefore useful to describe their most differentiated form. For their understanding, we begin with the fact, noted above, that all knowledge is relational. Its relationality involves the observer, the observed, and a mode of symbolic representation for communication. Therefore, all knowledge comprehension requires some mapping of observer frames and rules for translating knowledge from one frame into the other. The relationality between observer frames is a relationality of the second order: more complex reference frames admit of the valid existence of more specific ones to which they are superordinate. The advance of knowledge in science as well as in philosophy and other domains often consists of the integration of disparate frames of reference into superordinate frames. This results in a remapping of the knowledge already available, so that new relationships and patterns can emerge. Thus, knowing reference frames is crucial for understanding knowledge methodologies—and we use the term here in its most generic sense, not merely referring to scientific methodologies, but including also the methods of intellectual craftsmanship used by the historian, the lawyer, or any other knowledge producer.

Our emphasis on the relationality of knowledge and the need for the study of frames of reference does not imply that we consider knowledge in any of its various forms as irrational. Quite the contrary, we consider the sociological study of frames of reference an endeavor to discover the sociocultural aspects in the operation of rationality itself.

We can now turn to a tentative discussion of the components of reference frames. In using the term frame of reference, we are obviously trying to point to those elements that an inquirer takes as established, as anchor points, and the standards to which his/her inquiries are related. These form the context and the enabling ground within which inquiry proceeds. They include value commitments as to what it is important to know, and epistemological and methodological assumptions. As is well known, these may vary from the rationalist position that general first principles are the anchor points of truth in the empiricist position that only concrete observations can be relied on. Many more shadings exist, including assumptions about the manner in which one ought to proceed from an experiential grounding to conclusions. There are further schemas (categories) into which information can be tentatively ordered so that it can be initially received. The frame also includes an inquirer's preferred mode of explanation and theory, and especially the tests through which the inquirer guarantees or validates the knowledge that results from the inquiry.

There are several components or aspects of frames of reference that, while interdependent, are still capable of some independent variation. These include the establishment of an experiential base for knowledge, done through considering certain intellectual elements as primitives or *givens*. It is to these that a set of categories, taxonomies, and methods—more generally put, intellectual *operators* —is applied. Through them information is ordered, often into broad *master models* or metaphoric images of the domain under inquiry. Such master models or metaphors help orient the inquirer and guide the construction of theories. In this connection, we are speaking of a phenomenon akin to early versions of Kuhn's idea of "paradigms" in science, even though we feel that the study of reference frames affords a more complete picture of knowledge production and reception than is possible with the paradigm concept alone. We also include in the concept of frame of reference the *reality or truth tests* by which both the basic beginning points of the experiential base and the knowledge outcomes are validated. Such truth tests tend to become embedded in institutional contexts so that they can be used officially to guarantee the validity of knowledge arrived at.

Another aspect of frames of reference concerns their boundaries. Inquirers tend to assume that their domain of inquiry is bounded; they tend to have conceptions of alternative reference frames and their proper domains of inquiry. This is related to the fact that every reference frame, no matter how specialized and articulated, contains a limited set of rules for mapping alternative frames of reference. However, not every reference frame has the same capacity for interpreting alternative perspectives. Simple frames can be easily mapped into complex ones, but the reverse is not the case. For example, professions that enjoy a high degree of control over the application of their knowledge tend to be less articulate in translating their perspective into alternative frames than professions in the reverse situation. We would argue that the reference frames of professional American medicine are more limited in this regard than, say, those of social workers or sociologists, whose work consists in studying alternative reference frames. (The self-serving nature of this remark may be easily forgiven on the grounds of its obvious truth.) The translatability of reference frames is related to the degree of articulation of the frame of reference. This has two aspects. There is the aspect of codification, or the standardization of cognitive assumptions, procedures, and modes of explanation as occurred, for example, in experimental learning psychology some time ago. The other aspect of articulation is that of reflexivity: that is, the degree to which the user of a reference frame is aware of employing a specialized perspective to which alternatives are possible, as, for example, in the sociological specialization of ethnomethodology.

These are the major aspects of frames of reference that form the proximate context for inquiry and the reception of knowledge. The social validation of knowledge or the determination of reality in a certain regard occurs in intersubjective spaces within the context of shared frames of reference and through reality tests.

REALITY TESTS

Reality tests are relatively structured occasions in which some symbolic representation claiming the status of fact is subjected to scrutiny in terms of criteria of truth. They are essentially procedures for validating an experience or observation. They constitute occasions for ascertaining whether an idea is valid to the extent that action can be taken that may entail risk. Thus, reality tests always involve a testing of a cognition, a decision, and performance based on the decision. Such reality tests differ as functions of the frame of reference to which they relate, but they always involve components of subjective experience, object observation in some modality, interpersonal confirmation, and symbolic or rational consistency. Such tests are differentially distributed throughout the social structure and the probability of their use varies systematically with such a structured location. Indeed, working communities that share epistemic criteria—we will be discussing these *epistemic communities* shortly—

may well be distinguished by the types of reality tests they utilize. The kind of reality test employed in specialized, technical working communities reflects the basic criteria of judgement that define the community itself.

Perhaps the most obvious type of reality test is the empirical test that is formed in the diverse specialized communities of science. At the core of empirical reality testing is a set of institutionalized methodological procedures. They standardize interobserver reliability in obtaining empirical data about recurrent phenomena from which the investigation abstracts more general principles. Empirical reality tests represent the defining characteristic of the scientific frame of reference and are one of the central criteria of scientific activity. Such tests become strategic performances in these communities and are built into institutionalized structures of work flow and the exchange of communications.

By contrast, the pragmatic reality tests require the actor to check on the workability of an aspect of the knowledge underlying a performance. The validity of this knowledge or technique is evaluated in terms of the consequences flowing from it in action. Pragmatic reality tests demonstrate knowledge through successful performance; they do not require systematic relations to abstract principle. Thus, certain professionals may tend to demonstrate the validity of their esoteric expertise by successful virtuoso performances that reveal their capacity to attain desired results. The belief in the facticity of knowledge generated by such performances may not survive detailed empirical scrutiny. Pragmatic reality tests tend to be used in instances when the knowledge to be applied has not been or cannot be made explicit or systematically codified. Pragmatic reality tests have serious weaknesses: they may involve a biased sampling of observations and/or situations and they do not permit generalization or systematic criteria for confirmation. However, they also are powerful constraints on knowledge and knowledge use, especially in the domain of common sense. In fact, the verification of working knowledge through pragmatic reality tests can become cumulative and, to a limited degree, codified into bodies of knowledge that can then be officially certified, transmitted, and used. This is precisely the case in the history of practicing medicine, as well as other professions that apply technical knowledge to concrete problems. Indeed, the practitioner may well give preference to the accustomed pragmatic reality testing procedures over the more rigorous empirical tests employed by the scientists. The skeptical idea that something may be true in theory, but not work in practice, is behind this notion.

Certain derivatives of pragmatic reality tests used in professional communities are the various decision rules employed by them. In law, it is assumed that to convict one innocent man is worse than to let a guilty man escape. In medicine, by contrast, it is better to impute disease mistakenly and treat it than to find nothing and thus risk missing a disease. We will illustratively elaborate on this point later. These decision rules are essentially guiding presumptions concerning appropriate action in cases of uncertainty that are based on the cumulative results of numerous pragmatic reality tests throughout the history of such professions.

In the case of authoritative reality tests, a reality construct is accepted as valid because it is received from or approved by some authoritative source that is highly credible and certified as legitimate. In effect, authoritative reality tests determine the credibility of an assertion and stigmatize knowledge as valid truth in terms of the source or validator of the assertion. Such authoritative reality tests are particularly important as the basis for accepting legitimate political interpretations of significant historical landmark events. In addition to their obvious significance in this domain, authoritative reality tests take on great importance in bureaucratic structures. That is, in those specialized associations, bureaucratic authority vested in an office is a central criterion for establishing valid knowledge about reality.

This brief description of authoritative reality tests suggests further complexity that must be noted: different segments of a community may subscribe to different kinds of reality tests. In a large modern research institute, scientists typically rely on empirical and, secondarily, logical tests to assess the validity of knowledge and events. The administrative staff of the organization is likely to rely on authoritative reality tests in evaluating the legitimacy and validity of communications they receive. This raises complex problems of relations between the two areas.

It is not necessary for our objective that we enumerate all significant types of reality tests. It should be sufficient to mention a few selected types as a basis for comparison with those already discussed. Rational reality tests typically assess the formal symbolic structure of assertions and arguments in terms of the criterion of logical consistency. Because of this criterion, rational reality tests are often closely associated with empirical ones. Consensual reality tests assess the reality and reliability of knowledge or belief in terms of the criterion of consensual validation. Here what most people agree is real is taken as a validating criterion. Consensual reality tests together with pragmatic ones are probably at the center of common sense knowledge. Magical reality tests are quite similar to pragmatic ones in that they involve an assessment of the consequences of employing certain specific techniques in the course of some performance. In contrast, mystical reality tests evaluate the validity of reality and knowledge in terms of its origins: knowledge is only accepted as meaningful if it originates in some transcendental experience not accessible to everyone.

It is possible to summarize some of the foregoing considerations by presenting a crude and extremely tentative typology of reality tests based on three dimensions: first, time and information saving (nonlearning) versus time and information spending (learning); second, emphasizing technique and performance versus emphasizing knowledge and cognition; and third, evaluation of the knowledge or reality source ("Who says so?") versus evaluation of the consequences or effects of the knowledge ("What happened?" "Did it work?"). In terms of these three dimensions, the following typology is tentatively suggested for the reality tests that have been described so far.

TENTATIVE TYPOLOGY OF REALITY TESTS

	Technique and Performance		Knowledge and Cognition	
	Assess Sources	Assess Consequences	Assess Sources	Assess Consequences
Information saving (nonlearning)	Imitative Traditional	Magical	Authoritative	Consensual
Information spending (learning)	Mystical	Pragmatic	Rational	Empirical

The fact that reality tests are linked to frames of reference is clear; further, they are likely to become institutionalized in relation to the institutionalization of particular reference frames. In exploring this matter, we turn to the topic of epistemic communities.

EPISTEMIC COMMUNITIES

There are certain work communities for whom epistemic criteria, that is those concerned with the production and use of knowledge, have primacy over other interests or aspects of orientations. Such communities may be organized as formal associations—even though the coextensiveness of a formal association with an epistemic community is not at all the norm—or they may be structured as informal groups and communities of reference. We use the somewhat artificial sounding term epistemic community because the phenomenon we wish to point to differs in certain significant respects from other kinds of groups, be they primary or associational. The key element is the primacy of epistemic criteria for activities that involve the creation of new reality constructs or their application to situations of practice.

For example, science can be considered an epistemic community that institutionalizes rigorous epistemic criteria of theory formation and empirical testing, articulating these in a complex manner through "norms and counternorms" with the social structures of scientists. Mitroff (1974b), for example, shows how the norm of scientific universalism, which is based on the impersonal character of science, is supplemented by an opposing counternorm, that of personal commitment to points of view. He argues that the ultimate rationality of science as a collective enterprise is based on "sociological ambivalence" (Merton).

Robert Merton and Eleanor Barber (Merton, 1976) have presented and analyzed the concept of sociological ambivalence in some detail. The notion of sociological ambivalence is quite different from psychological ambivalence. Basically, the notion points to the importance of contradictory normative expectations in the social definition of one role or in the operation of an institution. Indeed, Merton and Barber write that they see the structure of social roles in a special way:

> From the perspective of sociological ambivalence, we see a social role as a dynamic organization of norms and counternorms, not as a combination of dominant attributes (such as effective neutrality or functional specificity). We propose that the major norms and the minor counternorms alternatively govern role behavior to produce ambivalence.
>
> *(Merton, 1976:17)*

Far from limiting the effectiveness of role performance through such contradictions, they claim that "only through such structures of norms and counternorms, we suggest, can the various functions of role be effectively discharged" (Ibid.:19). Merton's essay on "The Ambivalence of Scientists" (1973:32ff) is a particularly powerful illustration of the role ambivalence plays in the context of epistemic communities. For example, a scientist is certainly expected to make newfound knowledge available to the scientific public. But at the same time he/she is also expected to avoid rashly "rushing into print." Or one might think about the conflict between the premium on priority and the norm for scientific modesty. The manner in which norms and counternorms, in their dialectic alternation, are systematically interdependent is most complex. The perspective of sociological ambivalence opens up an understanding of the rationality of epistemic communities being sustained by norms and counternorms in a most complex, and sometimes conflicting, practice. For us this also implies that in the work communities we are discussing here, epistemic criteria are supported by structures internal to the community—often through norms and counternorms—so that there is necessarily an interpenetration between social arrangements and the cultural domain. Sociological studies that limit themselves entirely to the mapping of social structural arrangements therefore are of limited value in such contexts, even though they remain important. Full understandings emerge on the basis of an understanding of the interdependencies between cultural substance and social arrangements.

This view differs significantly from the treatment of science as a whole or of scientific disciplines and professions merely in terms of organizational analyses, mapping communication links or patterns of career advancement. That point of view uses modes of analysis and methods in principles applicable to any social structure without considering the special characteristics of knowledge-oriented work. One illustration of the important, but limited, contributions that can be made from this point of view is part of the work of Nicholas C. Mullins (1973).

The term *epistemic communities* thus designates those knowledge-oriented work communities in which cultural standards and social arrangements interpenetrate around a primary commitment to epistemic criteria in knowledge production and application.

In these terms, science is not the only epistemic community. Any special way of knowing, whose development and elaboration requires the establishment of an autonomous social space, will tend toward the structure of an epistemic community. In fact, certain esoteric knowledge traditions associated with various groups of serious astrologers must be considered epistemic communities, just as much as the officially recognized and discipline-based modern professions.

The establishment of a common frame of reference with shared epistemic criteria provides all members of such a community access to a consensually validated perspective for the construction of reality. The perspective required by the epistemic community must also be integrated into a sense of personal identity; this may vary from complete personal transformation through conversion to the establishment of a segmental role identity, which the individual adopts in the capacity of professional worker. The structures and strategies adopted in socializing new members into such a community are constrained to fit existing arrangements of control and the system of rewards. In fact, socialization into an epistemic community requires that the motivational relevance of the particular kind of knowledge used by the work group be established as a value in itself. When this occurs, intrinsic motivation can be effectively supplemented by extrinsic rewards. This means that power is an integral aspect of epistemic communities, as it is an integral component of all social interaction. However, exercises of power within epistemic communities must come to terms with the peculiar quality of power relations in these sociocultural domains. The primacy of the commitment to particular modes of reality construction, obviously, requires a certain mode of power legitimation. More than that, knowledge strategies may be used as components of power strategies as well. Since epistemic criteria define not only the nature of work within the community, but also the terms of its legitimation through application, the quality of models for knowledge production and knowledge application or use becomes a determinant and constraint for both the internal workings of the community and transactions across its boundary to publics or clients.

The experiential base of an epistemic community's reference frame may well be a major point differentiating epistemic communities. Certain experiential bases are oriented toward pragmatic or empirical tests and as such emphasize the equivalence of all qualified observers. This is the prevailing principle in science and in the science-based professions. However, in the case of experiential bases for esoteric knowledge, which may be mystical revelations or superior knowledge claims based on charismatic authority, observer equivalance is denied in principle. These two extreme possibilities relate to reality tests based on work and logic, on one side, and tests based on identity and authority commitments, on the other. Their relative predominance will leave a major imprint on the organization of work, socialization processes, and reward structure of an epistemic community. These are aspects significant for the internal structure of the community, but they are also related to how possibilities of knowledge use are viewed. Certainly, the epistemic requirements of technological, that is calculated, knowledge application in the Comtian vein differ from those that require transformation of knowledge into requirements for disciplined professional practice. These, in turn, differ from those conceptions of knowledge use that demand transformations of consciousness in order to produce proper action—as the Marxian view recommends. The manner in which an epistemic community orients itself to its internal criteria of epistemic validity, and thus defines the role of the observer or knowledge producer, becomes articulated in complex ways with the range of possible models for knowledge use and *their* epistemic standards.

Both elements, the internal epistemic structures and the conception of permissible ranges for knowledge use, enter into the analysis of the social and cultural autonomy of epistemic communities. Autonomy, at least to the extent of permitting the work to progress on its own terms, is not incompatible with control in other regards—as the continued relatively autonomous existence of science under totalitarian regimes has shown. However, *some* autonomy is required for the specialized frame of reference and work orientation of an epistemic community to unfold. In turn, such specialization results in the establishment of rationales for knowledge structuring, specialized bodies of knowledge, and the construction of specialized and sometimes highly esoteric languages—that is, the development of a cultural domain on its own terms. Such autonomy may be legitimated through the services of knowledge production and granted in a fiduciary manner; that is, the authority system of society at large yields, on trust, a cognitive monopoly to an epistemic community—as is the case with academic disciplines and the knowledge-based professions. On the other hand, autonomy may be established surreptitiously by going underground, as it were, and establishing in various modes domains for the creation and preservation of secret knowledge. Whatever the modes of autonomy, they will leave a profound imprint not only on the social organization of the workers in an epistemic community, but on their cultural product as well.

CULTURAL FREE SPACES

Cultural free spaces are contexts or domains in which otherwise effective restraints and authoritative strictures are at least partially suspended, lessening controls on thought and action. They are important, but by no means the only or necessarily the dominant sources of innovation. Scholarly and artistic patronage in Renaissance courts serves as one example of such free spaces. The effervescent café society of Vienna is another. Indeed, the origins of science in amateur activity clearly link it to such free spaces. These interstitial domains in which, above all, the restrictions of authority and status are lessened take many different forms. They may arise in the encounters of marginal men during a mingling of civilizations (as in the role of the Jesuits in China), in the shelter of the great cities, or on the fringes of the great social movements (but never in their center where authority is wielded with utmost seriousness). In spite of the diffuseness of their forms, cultural free spaces are phenomena of the utmost significance for the understanding of the knowledge system. These phenomena have in common the suspension of criteria of authority and status, and the suspension of at least some aspects of the taken-for-granted structures of common sense.

For example, Goethe reports in his account of the campaign of 1792 that during the bombardment of Verdun by the allied forces, in the midst of battle, he conversed for many hours with a great prince about the phenomenology of colors—a subject brought to mind by the spectral refraction of light in a fish pond. This slightly bizarre episode illustrates some further points about cultural free spaces: their boundaries are set by frames of reference, or states of mind. And yet, there is a massive social structural reality behind the phenomenon of cultural free spaces supporting their existence. The young American Peace Corps volunteer may, indeed, enjoy the benefit of a limited and temporary cultural free space in the host country, being able to invite cabinet ministers and peasants equally to parties. Status outside the system provides the opportunity for this. Yet it is clear that both young Goethe's access to the prince, and the prince's interest in him, and the role of the Peace Corps volunteer are backed by powerful social structural forces. Within the cultural free space, playful exploration and examination of alternatives can take place. This is what Moore calls "autotelic environments" (A. A. Anderson and O. K. Moore, 1969).

Cultural free spaces must be accomplished by a structural moratorium that releases their participants from at least some aspects of adult role status relationships, obligations, and responsibilities. It is this moratorium that encourages experimentation and innovation with alternative interpretations and behaviors. Conventional prescriptions and proscriptions that channel the behavior of normal adults are lifted, along with conventional cultural constraints on symbolic innovation, so that the exploration of alternatives becomes feasible. Cultural free spaces are essentially secular phenomena—they are predicated on the selective seeking of change and the rewarding of innovation. Indeed, in spite of the suspension of status and authority (to some extent), there normally remains an element of competitiveness that may well take the form of competing for the most imaginative innovation.

Let us note briefly that cultural free spaces and epistemic communities may, but do not necessarily, overlap. Indeed, amateur science had all of the characteristics of a cultural free space, but few would argue that the professionalized and bureaucratized working environment of a large industrial research and development laboratory shares these characteristics. After all, professionalization has traditionally brought with it an aura of seriousness and respectability that frowns on unconventional thought and behavior, and it is precisely unconventional interpretations and actions that are required in order to generate the innovative excitement of cultural free spaces. It is because of such circumstances that the term *Bohemia*, with all its different connotations, came to be associated with many, but by no means all, of the significant cultural free spaces around the end of the nineteenth century.

Bohemians evolved a relatively consistent system of ideas, which enabled and encouraged symbolic experimentation and cultural innovation, free from the constraints of institutional restrictions and common sense. These ideas included, for example, the notion of salvation by the child and a cult of childhood. This idea referred to the special potentialities everyone has at birth and their gradual erosion and destruction by a standardized society with mechanical modes of teaching and socialization. It also idealized a childlike innocence, common sense, and an ability to perceive alternatives other than those prescribed by established adult society. Another notion characteristic of Bohemian contexts was that of self-expression and the realization of individualities through creativity. This also encouraged symbolic innovation and experimentation. Indeed, experimentation may include concerns with mind expansion, alternate states of awareness, and hallucinations as keys to mystical experience. These were prominent in the generalized, romantic love of the exotic.

Cultural free spaces in the interstitial and marginal surroundings of great social movements are another example. Socialist movements especially generated an attraction for the young and for intellectuals who wished to create free spaces, explore alternatives, and debate possibilities. In an attempt to describe ideological groups among the small revolutionary circles that emerged in Russia during the late nineteenth century, Nahirny observed that (1963:398):

> ... there are distinct social groups which are qualitatively different from personal and functional ones. These are groups which make their members relate to one another in the light of ideas and beliefs rather than in the light of personal qualities or specific functions which they perform. . . . It is one of the most striking and general features of ideological groups that they frown upon and oppose vehemently any display of personal affective attachments among their members. Ideally, members of ideological groups are expected to avoid familiarity and to forego willingly all familial attachments and personal ties for the sake of some such central symbols and collectivities as socialism, revolution, people, or proletariat.

These contrast with ideological informal groups that Nahirny (Ibid: 404) describes and that are linked to cultural free spaces.

> Ideological formal groups do not admit . . . either the separation of public and private spheres of life, or the clearcut segregation

of the individual's roles. Their members do not participate and are merely in the capacity of one narrowly defined and functionally specific role of "official." They continue to demand, like ideological informal groups, a total commitment to the cause which is now authoritatively defined and institutionalized. It is here that they differ strikingly from ideological informal groups. Members of ideological informal groups are recruited on the basis of individual contacts and mutual confidence, members of ideological formal groups on the basis of formal requirement and regulations. The former groups are held together by "inner convictions" of their members, the latter by the organizational structure that has come to embody them. By establishing, as it were, an "immediate rapport" with their ideas, members of ideological informal groups are inwardly compelled to conform with them. Members of ideological formal groups . . . conform primarily to organizational principles.

Such ideological informal groups have been a characteristic of much movement activity. They did arise often in cultural free spaces and fostered intellectual and interpersonal experimentation. Literary movements, youth movements, feminist movements, and cultural movements of various kinds readily illustrate this point.

CONCLUSION

In this chapter, we have sketched our perspective of the social construction of reality, moving in our presentation of concepts from the experience of reality in consciousness, the subjective domain, to its anchorage in social structure. Much remains to be said about the relation between social structure and knowledge, and it is this topic we address in the following chapter.

5

Social Structure and
Knowledge

THE IDEA OF SOCIAL STRUCTURE

A consideration of certain principles in the relationship between knowledge and social structure is now necessary. Our purpose in this chapter is quite limited. We do not intend to present a theory of social structure, nor to debate all the profound sociological issues connected with it. We merely introduce our conception and analytical imagery of social structure in order to present certain necessary tools for the subsequent, more specialized treatment of knowledge production and use systems. Our discussion of social structure therefore can be brief, omitting much of what is generally treated under this heading in sociological theory.

After a general discussion of basic concepts, we turn to the issues that emerge when one views social structures as frameworks for knowledge, particularly bodies of knowledge. Knowledge of society or, more specifically, knowledge of social structure plays a special role in social life through its reflectivity. The relationship between social structure and knowledge is a complex and dialectical one. After considering this theme, we conclude with some rather detailed illustrations of relations between structural arrangements and cognitive systems that exemplify the abstract conceptions we have presented this far.

The imagery of social structure is central to all sociological analysis. By social structure we mean a conception of society as a relatively stable pattern of relationships among social positions, roles, groups, and institutions. The term *structure* means a set of relationships among the parts of a totality; it is the organization of a whole. Structure, then, is a pattern of relationships within a system or whole and it must be discussed in the framework of such systems.

To perceive structures requires specialized attention and a particular point of view. After all, living and complex wholes may be organized in the most intricate and overlapping ways. The living organism, for example, could be described as structured into many patterns, depending on one's focus of attention. These kinds of considerations have raised a false issue: structure as an attribute of someone's theory versus structure as inherent in reality itself. Another problem is the integration and the static or dynamic character of the social wholes of which structure is an attribute. A brief digression will clarify these matters.

In a simplified way of speaking, the relationships among parts that make up a whole may be of at least three different types. If all parts have the same attributes and are thus identical with each other, then the totality simply shares these attributes—as in a collection of green objects. Secondly, a whole may have a cognitive structure, as in the case of a logical argument where the logical relationships structure the whole. This type of whole-part relationship emerges as particularly important in cultural objects, as for example in the structure of ideologies. Such wholes are structured as logico-meaningful systems in Sorokin's sense. Finally, while the parts may not necessarily share any of the attributes of the whole, their interaction may give rise to new emergent characteristics of the totality that are not reducible to the properties of the constituent parts, as for example in a collection of red and yellow objects that may appear orange from a distance. Here we are dealing with a relationship that, in the ideal case, can be described only by a theory explaining the attributes of the whole as a function of the interrelations among the attributes of the parts. Social structure is usually of this type.

The study of social structure requires a theoretical focus on sets of social relationships in such a manner that the dynamic qualities of social entities become understandable. In this sense, social

structure is always represented in terms of someone's theory or
proto-theory, however descriptive of social reality it is. If the relationships among the components of social structure are conceived of
as dynamic, structure emerges not as a static conception of social
arrangements but as defining pathways of change in a social system.

When sociologists study social structure, they select certain
dimensions that permit the description of a system of social relations. At least in this sense, it is the sociologist's perspective from
which social structure is mapped. Blau has proposed that "social
structure refers to the differentiated interrelated parts in a collectivity, not to theories about them. . ." (Blau 1974:615). He asserts that,
by contrast, social structures often have been viewed as if they were
entirely constructs in the mind of the analyst. An illustration of this
position is taken from Lévi-Strauss (1963:279):

> Social relations consist of the raw material out of which the
> models making up the social structure are built, while social
> structure can, by no means, be reduced to the ensemble of social
> relations to be described in society.

Blau emphasizes that the dimensions and parameters of social structure are inherent in social reality itself; Lévi-Strauss seems to emphasize the constructive aspect of sociological theorizing.

In a basic sense this particular disagreement is specious. There
is no scientific observation that is not conceptually and, at least in
this rudimentary way, theoretically guided; hence, Blau's accounts of
social structure also reflect his theoretical orientation. At the same
time, such accounts are constrained by empirical measurements and
the characteristics of the reality to be mapped. This raises an important peculiarity of all concerns with social structure. Social structure
is an arrangement of relations among conscious, acting persons.
These persons have some symbolic representation of themselves,
their groups, and of their social structure. To be sure, the analyst will
not take at face value and as ultimate truth what members of a
society believe about themselves and their society. However, one
must deal with what they take to be their factual knowledge and,
especially, their factual social knowledge.

SOCIAL STRUCTURE AS A
KNOWLEDGE-CONTAINING SYSTEM

Any account of the organization of social entities must deal with the fact that they are already knowledge-containing systems and, furthermore, that they are capable of learning. That is, their members construct their own social structure, in part through their symbolic definitions of it. The members of a society find themselves arranged into social roles, positions, and relationships, and they give symbolic accounts of them. A sociological study of social structure, therefore, must build on the implicit and explicit symbolizations and folk accounts of social relations found in society. We follow here the conception advocated by S. F. Nadel who, building on the work of Talcott Parsons, defines social structure as follows (1957:12):

> We arrive at the structure of a society through abstracting from the concrete population and its behavior a pattern or network (or "system") of relationships obtaining "between actors in their capacity of playing roles relative to one another."

> *(The quote within this quotation*
> *refers to Parsons, 1949:34).*

This conception articulates an image of society as differentiated into symbolically defined roles that are linked to each other by social relationships to establish a network or web. It is important to emphasize that such a positional network is arranged largely in terms of symbolic designs and is already a knowledge-containing system. The observer who wishes to prepare an account of it is bound both by personal theoretical interest and the inherent characteristics of the symbolic design of that knowledge-containing system.

We may now consider two rather different, but supplementary, analytical aspects of social structure that emerge from different perspectives of the observer. One of these aspects comes into view when we understand society from the point of view of a conscious, participating member who observes it. Within this frame of reference, a society appears as a large network of purposeful actors, each of whom orients self to others in the context of a meaning system. Each of these actors has a particular orientation and point of view. An account of this image of society results in a description of the distribution of orientations.

At the same time, a rather different perspective is possible from which a detached viewer may see society as consisting of interdependent, behaving objects that relate in a describable fashion to each other. Here social structure is viewed from an ecological reference frame, as a setting within which relations among actor-objects exist. We will call this aspect the situational structure of society; in this perspective, society is seen as a vast distribution of situations and situational constraints.

The orientational structure of a society includes such phenomena as the images its members have of it, the distribution of beliefs about social facts, and the knowledge and evaluations of groups or institutions. In this perspective, members are seen as acting subjects who orient themselves through their reference frames to norms, including legal standards, that influence their activity. Social structure in this view hangs together through relations of trust and legitimacy: it is divided by social boundaries and ripped by conflicts. In the alternative view, society is seen as a distribution of persons in environments or settings. One can observe that activities occur in certain specific times and places and are linked to the activities of others in a systematic and regular fashion. From this point of view there emerges a picture of society as a distribution of settings or situations. Viewed as a situational structure, society is an ecology of objects that move from one setting to another. We can treat these two complementary, analytical views of social structure as representing the social arrangement for the distribution of orientations and the pattern in the allocation of situations.

Both of these aspects are, of course, interdependent. Situations are meaningful to actors in terms of the orientations they bring to them, but they are rarely, if ever, completely defined only by the participating actors. Situations are environments as settings for action; their meaning depends on general cultural definitions. We enter situations not so much attempting to impose our definition on them, as trying to decipher the meaning of the situation. In general, we can say that the ecology of settings (situations) is shaped by the character of the natural environment, the nature of the division of labor and social differentiation, and the availability of power that can be brought to bear on them. It is clear that the situational structure of society depends on the nature of power and its relation to environments in that society. It is essentially the social structure of power that results in a specific allocation of situations to persons. Following this line of thought, the situational structure implies a conception of social stratification in which individuals and groups attempt to control and stabilize the situations with which they deal. Strategies for situation control, in relation to the actual distribution of situations and their demands for action, thus become one important attribute of social structure.

The conception of an orientational structure of society implies the notion that the perspectives of actors are far from being randomly distributed. Every observer of society is struck by the fact that there are limited symbolic and cultural horizons among its people. Patterns of socialization, reflecting differentiations among groups, institutions, roles, and social positions, demand the adoption of relatively specific points of view. In this connection, we see society as cognitively differentiated, whereas the emphasis on the situational structure stresses the ecological differentiation.

The two aspects of social structure have their common origin in cultural dimensions of structuration inherent in the activities of a society's members. The situational structure of society, while conceived of analytically as a distribution of settings, cannot truly be thought of independently of the manner in which such situations are culturally defined. Thus, the sociological understanding of social structure in both its major aspects is only possible in relation to some understanding of the symbolic dimensions through which actors organize their relations to environments and each other. These can be thought of as dimensions of the orientation systems in terms of which actors assess situations, people, and groups; however, they do not necessarily inhere in any given actor's orientation. Powerful systems of social control within each society have established structural arrangements with specific cultural meanings, which are enforced through positive and negative sanctions.

The manner in which actors categorize themselves and each other reflects either implicit and intuitive or explicit and occasionally systematic social measurements. We now can think of social structure as providing for the assessment, through measurement, of individuals, their orientations, and their actions in relation to the situations they face. Viewed in this fashion, social structure demands that individuals and collectivities locate themselves in relation to each other and symbolize their identity. In this sense, there is an aspect of reflexivity in social structure. It is the crystallized residue of structuring processes. Thus, we must emphasize that the distinction between orientational and situational aspects of social structure is not absolute. The former emphasizes the cognitive and active aspect, the latter the power and constraining aspect inherent in the web of social relations.

We have touched on some intricate matters, but the basic imagery of social structure emerges as a relatively simple one. We see society as a network of relations within which actors acquire and use differentiated orientations and that, at the same time, regulates (but rarely controls completely) the flow of situations with which the actors must deal.

SOCIAL STRUCTURE AS A FRAMEWORK
FOR KNOWLEDGE

Our description of the orientational structure of society emphasized that all social structure has a cognitive aspect. We now turn to the question of how social differentiation, in particular social stratification and institutionalized inequality, provides socially structured frameworks for differentiated knowledge. The particular sociohistorical and cultural location of a group obviously influences its cognitive perspective and available knowledge. The way in which social arrangements become frameworks for knowledge raises more specific questions. Several of these have been discussed by Gurvitch in his work *The Social Frameworks of Knowledge*. Gurvitch means by sociology of knowledge (1972:16):

> The study of functional correlations which can be established between the different types, the differently emphasized forms within these types, the different systems . . . of knowledge and, on the other hand, the social frameworks such as global societies, social classes, particular groupings and various manifestations of sociality. . . .

By social frameworks Gurvitch means solidarities, groups, social classes, and total societies that evolve characteristic priorities and strategies for knowing. For example, Gurvitch describes the way in

which small scale local groups develop a mode of knowing characteristic of themselves. Small scale local communities such as hamlets or
villages face, among other things, the task of keeping order, controlling neighborhood relations, carrying out economic activity, and so on.

> Leaving aside the influence of social classes and global societies,
> small scale local groups are sources of a specific perceptual
> knowledge of the external world. The space in which the world
> is placed is primarily egocentric and concentric: that of flower
> and vegetable gardens, fields, forests, road and communication
> routes with neighboring villages and small towns where markets
> are held. The external world tends to expand but, at the same
> time, it risks being placed in a diffuse space when it concerns re-
> lations with the large towns, the state and its capital (where the
> centers of administrative and political organization are).
>
> *(Ibid.)*

These small scale groups, forming in a circumscribed locality, are the
sources of a common sense knowledge adapted to their style of life.
They also provide sources of technical knowledge. Thus, in many
European hamlets, technical knowledge concerning agriculture is of
the traditional variety and is transmitted as an integral function of
the local community structure. Gurvitch also describes a particular
type of political knowledge characteristic of these groups.
    Another aspect of Gurvitch's approach is his discussion of
factories and workshops as social frameworks of knowledge. In these
kinds of groupings, technical knowledge predominates; it is adapted
to the machinery and the production process itself. Immediately following it in importance is a particular kind of political knowledge,
"understood primarily as tactics of adaptation to the moods of team-
mates, foremen, or engineers" (Ibid.:71). The perceptual knowledge
of the external world also can be described as having a particular
structure in the factory.

> It concerns the external world of the workshop, the factory, and
> the enterprise. Its objects are the premises, the machines, the of-
> fices—including those of the directors and engineers—which are
> so near and yet so far from the workers; the concrete space and
> specific time in which the work and life of the members of the

workteam occur, whether or not they work on assembly line production; and finally, the distances to be covered by the workers
to meet their bosses and to get to their machines, and the time
taken up in traveling from home to work.

(Ibid.:72)

With this description Gurvitch is trying to characterize a particular
time-space structure that derives from the social framework of the
factory and becomes the context for knowing specific information
about other persons and objects.

Using this general approach, Gurvitch proceeds to offer
sketches of the cognitive systems that characteristically arise in differently structured social settings. Particular groups, entire social
classes, and whole societies can be characterized in this manner. Gurvitch's approach is to take a basically ecological or situational differentiation model of society as his point of departure in order to describe the orientational structures and cognitive systems arising in
such contexts.

The approach does not explain how systematic and organized
knowledge (bodies of knowledge) are anchored in social frameworks.
We touched on the matter in the earlier reference to the idea of epistemic communities. These are communities in which particular
knowledge-related criteria take precedence over other criteria of
judgment. Often epistemic communities are specialized working communities, such as technical or professional specialists. The location of
such communities in the social structure has much to do with the
manner in which the community can achieve autonomy and maintain
its particular frame of reference.

ILLUSTRATIONS: SPECIALIZED EPISTEMIC COMMUNITIES AS STRUCTURAL FRAMEWORKS FOR TECHNICAL KNOWLEDGE

We now turn to illustrations—somewhat arbitrarily selected—
of these abstract matters. Our examples are drawn from the world of
contemporary professions, particularly in the medical area.

Identity as an Epistemic Criterion:
The Case of Psychoanalysis

This illustration is an example of intensive, advanced socialization into a particular specialized, technical working community; the case is that of the community of professional psychoanalysis.

Psychoanalysis occupies the status of a medical subspecialty. This medical status is a result of decisions taken in the 1920s by the International Psychoanalytic Association under extreme pressure from the American Psychoanalytic Association to restrict the practice of psychoanalytic psychotherapy to medical professionals. Thus, none of the sixteen A.M.A. and A.P.A. approved psychoanalytic institutes accept nonmedical candidates—with the exception of an occasional research candidate who participates in the training in order to acquire additional theoretical and/or research skills. It has also meant that many noninstitute certified psychiatrists and clinical psychologists, who actually practice psychoanalytic therapy, have to refer to themselves as analytically oriented or lay analysts. However, there are numerous other types of postgraduate psychoanalytic institutes, which do not require a medical background, where aspiring psychotherapists can be initiated into the esoteric art and science of psychoanalysis. These range from the prestigious William Allenson White Institutes—of which there are three—that have a more Sullivanian interpersonal psychiatric orientation, but accept qualified applicants with Ph.D., M.S.W., and LL.B. degrees in addition to medical psychiatrists, to the National Institute of Psychoanalysis in New York, which does not have any rigid policy concerning prior educational certification in keeping with the views of its founder, Theodore Reik.

The result of all these divergent paths to psychoanalytic practice is that the most prestigious as well as cohesive group—the orthodox defenders of the true faith—consists of analysts who are medical specialty, board certified psychiatrists, and it is this group that we describe in some detail. Only this group of technical, specialized professionals are legally entitled to refer to themselves as psychoanalysts despite the fact that numerous other psychotherapists in fact practice psychoanalysis. Thus, we use the term psychoanalyst to refer to psychiatrists who have gone through one of the approved A.M.A. or American Psychoanalytic Association psychoanalytic institutes. The discussion will focus on three psychoanalytic communities that have received systematic empirical attention: specifically, those in Chicago, Los Angeles, and New York City.

A former President of the American Psychiatric Association observed:

> The only true specialty inside the general field of psychiatry is psychoanalysis. It has a body of knowledge, criteria for selection and training of its candidates, institutes to carry out training and a method of certification.
>
> *(Blain, 1953)*

It is unimportant whether that assertion continues to characterize psychoanalysis in relation to its psychiatric parent profession; the important point is that psychoanalysts comprise a very small, highly cohesive and homogeneous elite within psychiatry.

Within each major American metropolis, the psychoanalytic institute constitutes a well-defined epistemic community of professional practitioners. The number of institute members, both certified practitioners and analytic candidates (who are still in training), is remarkably small; certainly small enough to constitute a cohesive community in every sense. Specifically, in 1966, there were only 168 certified psychoanalysts in Chicago, another 241 in Los Angeles, and even in New York City there were only 761 analysts—although these figures do not include analytic candidates in the respective cities. (See J. Marx, 1969; J. Marx and Spray, 1972; and Henry, Sims, and Spray, 1971:204 for discussions of these figures.) Part of the reason why there are so few professional psychoanalysts in these huge cities and throughout the country lies in the extensive training that must be completed before one can assume the title of psychoanalyst.

The road to becoming a professional psychoanalyst begins with completion of the baccalaureate degree whereupon the future practitioner must embark on a four-year program in medical school, followed in many cases by a one-year internship (although this is no longer necessary) prior to beginning a three-year psychiatric residency at some hospital closely associated with an analytic institute. Around the beginning of the third year of residency, applications may be submitted to both the local and other psychoanalytic institutes. If the applicant survives the intensive interviews and screening at the institute, and passes the psychiatric specialty boards, he/she then begins specific psychoanalytic training, which generally takes another five to eight years, depending largely on the length of the training analysis (Lewin and Ross, 1960; Henry, Sims, and Spray, 1971 and 1973).

The assertion that the communities of professional psychoanalysts in the three cities are highly cohesive and homogeneous is supported by the fact that 62.1 percent of the members describe themselves as having a Jewish cultural affinity, in contrast to 16.7 percent who report a Protestant cultural orientation, 2.6 percent a Catholic cultural orientation, and 18.6 percent who reported having no orientation to any particular cultural tradition. Moreover, comparing psychoanalysts with nonanalytic psychiatrists, clinical psychologists, and psychiatric social workers reveals that "psychoanalysts are much more likely to be Jewish or to claim no cultural affinity and less likely to have either a Protestant or Catholic affinity than is true for the other three professions." (Henry, Sims, and Spray, 1971:10). Thus, psychoanalysts have a level of occupational-cultural homogeneity far higher than that of any of the other core mental health treatment professions. Moreover, although all four types of mental health professionals come from metropolitan community family backgrounds, this was true of 74 percent of the analysts, making them the most metropolitan group in terms of community background.

Another dimension of occupational homogeneity in which psychoanalysts exceed the other mental health professions involves ethnicity—the countries to which professional therapists traced their national origins. Although eastern European ethnic backgrounds were most frequent in all four mental health treatment specialties, a larger proportion of psychoanalysts had eastern European ethnic origins than any of the other groups. (See Henry, Sims, and Spray, 1971:12-14, for the data on urbanism and ethnicity among the four professional mental health treatment groups in the three cities.)

These and numerous other findings (concerning past and present religious beliefs and affiliations, political orientations, social class backgrounds, etc.) leave no doubt as to the sociocultural homogeneity of psychoanalysts, which far exceeds that of the other, comparable professional treatment occupations in the mental health field. Thus, recruitment into psychoanalysis is powerfully influenced by sociocultural background. This means that the major distinction

within the mental health field is between those trained as psychoanalysts and those professionals trained generally as psychotherapists.
Since psychoanalysts as a group rank highest in occupational prestige
—both within and without the mental health field—and constitute
the most professionalized group of psychotherapists, we conclude
that advanced, specialized training and socialization do not reduce
the influence of traditional, background sociocultural ties and affiliations. On the contrary, there seems to be a marked congruence between specialized professional identities acquired through intensive
socialization and sociocultural traditions that stem from family
origins.

The Study of Careers in Mental Health Research found psychoanalysts to be far more homogeneous than any of the other three
mental health treatment specialties with respect to a wide array of
professional functions, roles, assumptions, ideologies, and work-
setting variables (Henry, Sims, and Spray, 1971 and 1973; J. Marx
and Spray, 1970 and 1972). For example, with respect to the variable
that is of critical importance for our purposes, Henry, Sims, and
Spray observe that (1973:57):

> The four professional groups vary considerably in the extent to
> which they are ideologically homogeneous. As might be expected,
> psychoanalysts vary least in their ideological positions: nearly
> two-thirds of them are advocates of the pure . . . (that is, clas-
> sical-orthodox) . . . psychoanalytic ideology, with the over-
> whelming majority of the remaining analysts claiming the psy-
> choanalytic position as their major therapeutic orientation.
> There is considerably greater ideological variance among mem-
> bers of the three remaining professional groups, although the
> popularity of some version of the psychoanalytic orientation is
> pronounced in each of them.

It is possible to approach the analysis of the psychoanalytic
epistemic community from a different perspective: fully 57 percent
of the 3,419 mental health practitioners in the four core treatment
professions, in the three cities, designated their major therapeutic
orientation as psychoanalytic—and no other ideological position
secured allegiance from more than 14 percent of the entire sample.
When data on respondents' other possible additional therapeutic
orientations were examined, 53 percent of those who listed psycho-

analytic as their major orientation claimed they had no additional orientation. This contrasts with most respondents who claimed that their major orientation was something other than psychoanalytic; the vast majority of these respondents also listed at least one additional orientation that differed significantly from their major orientation. This means that the psychoanalytic frame of reference and treatment ideology itself is uniquely homogeneous and "pure" of ideological contamination. But the most remarkable finding is that the most popular and prestigious therapeutic frame of reference is also the most pure and exclusive—rather than the most ideologically vaguely defined and hence contaminated by other, unrelated therapeutic orientations. As Henry, Sims and Spray note (Ibid.):

> The ideological orientation demonstrating the greatest capacity for eliciting exclusive commitment is, of course, the psychoanalytic. For each of the professional groups, practitioners who designate their therapeutic orientation as exclusively psychoanalytic manifest a pronounced tendency to specialize in the performance of one-to-one psychotherapy.... More important, perhaps, is the fact that the pure psychoanalytic ideology is clearly segregated from other ideologies in all professions but social work.

The affinity expressed by psychoanalytically oriented practitioners for the performance of individual psychotherapy in private practice settings warrants consideration because it provides an important clue to the ways in which a shared frame of reference facilitates the development of attachment to particular professional role structures. The performance of one-to-one psychoanalytic therapy requires a unique and highly specialized set of skills in social interaction. More importantly, the application of these specialized skills occurs in a private, intimate, highly variable situation that is, by its very nature, fraught with uncertainty for the psychotherapist. Under such circumstances, it is plausible to expect practitioners to be attracted to a frame of reference that has been explicitly and systematically formulated. That is, unlike competing frames of reference, the psychoanalytic one has a theoretical fountainhead, and has been subjected to numerous attempts to systematize and codify basic premises, approaches, and techniques. There is also a set of standard sources and references, which serve as basic theoretical and methodological guidelines for that frame of reference. Thus, we are suggesting that exclusive adherence to a systematically formulated and codified, as well as uncontaminated, reference frame facilitates commitment to a specialized, technical, professional role and epistemic community of workers.

In focusing on the epistemic community of professional psychoanalysts and the psychoanalytic frame of reference up to this point, we have emphasized the sociocultural and ideological homogeneity that characterizes analysts, as well as the doctrinal exclusiveness and purity of the psychoanalytic reference frame. We now take up the critical factor that differentiates psychoanalysts and their frame of reference from other psychotherapeutic specialists and reference frames, and that is largely responsible for the remarkable professional and doctrinal stability and continuity of psychoanalysis; namely, analytic institutes. Although the mean length of time in training for psychoanalysts, from undergraduate education to the completion of postresidency training, is 15.1 years, the last part of this training, at a psychoanalytic institute, was reported to be the most significant stage in their professional preparation. For about 16 percent of the analysts in the three cities, this last stage took three years or less; 58 percent took from four to six years to obtain institute certification; and 26 percent took seven years or more of training and socialization at an institute (Henry, Sims, and Spray, 1971: 142). Moreover, "to the extent that trends in the profession can be inferred from age grouping, the evidence suggests that length of psychoanalytic training may be increasing" (Ibid:141).

Psychoanalytic institutes involve the future analyst in extensive courses and case conferences, considerable clinical work under the supervision of experienced analysts, and personal psychotherapy. The "Careers Study" found that analysts evaluated their clinical experiences at the institute more positively than their course work, although both types of experiences were enthusiastically endorsed as useful by an overwhelming majority of analysts. This is understandable: by the time the mental health trainee reaches this point in training, processes of mutual selection by trainee interest and institute requirements probably eliminates those who would not find the experience wholly profitable. Given their already very respectable credentials for practicing psychotherapy, the fact that they elected to pursue further training suggests that the anticipated rewards in both skill and prestige were quite high. The nearly universal satisfaction with the institute program suggests that the expectations were indeed met.

However, there was almost unanimity of opinion among the analysts studied that the single most important experience in their entire preparation for their professional career was their training analysis at the institute. This uniquely intense socialization process—involving changes in attitudes, values, motivations, and identifications—explicitly has both personal and professional objectives. Thus, discussing professional socialization generally, Levinson observes (1967:258): "The socializing experience brings about changes in certain personal characteristics; these affect the student's subsequent career and are in turn affected by it." One crude index of the importance of the training analysis to the overall program offered to analytic candidates is that the total mean time taken by the candidates' personal analysis (in the Careers Study sample) is 5.3 years and the total mean length of the entire period of training at an institute in that study is 5.4 years (Henry, Sims, and Spray, 1971:142, 171).

We now are in a position to see the diverse ways in which psychoanalytic institutes socialize prospective members into the epistemic community of professional psychoanalysis and reconstruct their identity. First, there is the cohesiveness of psychoanalytic theory itself: its relatively high degree of impermeability, explicit codification, comprehensiveness, and lack of contamination by other viewpoints. Moreover, it is a self-validating body of knowledge—in that one can employ psychoanalytic theory to explain either of the opposite reactions to the same stimulus. It tends to be validated in terms of the success of treatment techniques based on it, and evaluations of the outcome of therapy are notoriously unreliable as well as confounded by the high probability of spontaneous remission of the disorder after several months.

The second component in the psychoanalytic community's socialization armamentarium involves a mutual process of selective attraction to and by the institute in terms of sociocultural background characteristics. This should not be surprising. In the everyday social world, "birds of a feather flock together," presumably partly because they speak the same language and can feel comfortable not only with the terminology used, but also with the cognitive and intellectual processes and assumptions used in deriving conclusions. Since psychoanalysis is an exclusively verbal therapy, it is not surprising that institute selection committees should admit applicants who can speak their own language. However, psychoanalytic encounters are unusual by virtue of their peculiar emphasis on affective communication. Since effective communication takes on such strategic significance to members of the professional psychoanalytic work community, such interpersonal factors as intellectual abilities and cognitive styles are viewed as necessary, but not sufficient, conditions for accepting an applicant to an institute. What is needed, in addition, are certain feeling states or emotional predispositions that the institute regards as appropriate for becoming an analyst. Since all applicants possess the requisite technical attributes for admission—medical background and psychiatric residency certification—selection is made on other grounds.

The Marx and Spray (1972) analysis of the Careers Study data suggests that institute selection committees use social class characteristics of applicants to gauge their intellectual and interpersonal suitability for the performance of psychoanalysis, and use religiocultural factors to assess their emotional and introspective suitability for joining the community of psychoanalysis. Insofar as psychoanalysis is primarily an affective emotional process rather than an intellectual-cognitive therapeutic approach, it is understandable that analytic institutes select candidates who are homogeneous in terms of cultural background more than in terms of social class origins. In other words, institutes utilize religiocultural congruence between applicants and selection committee members to bring about a selective matching of emotional factors between present and future members of the psychoanalytic epistemic community. Thus, institutes attempt to attract and admit candidates who have the emotional prerequisites for the specialized, technical work role of professional psychoanalyst by selectively recruiting, at least in part, in terms of ethnic and religiocultural considerations, which produces the homogeneity of member family origins noted earlier.

Once at the institute, the candidate is exposed to three processes designed to make him/her a member of the epistemic community: professionalization, indoctrination, and resocialization. In the course of performing psychoanalysis while being closely supervised by members of the institute faculty, the candidate acquires the specialized techniques and technical skills that constitute psychoanalytic expertise. Thus, clinical experience under supervision is the basis of professionalization processes at analytic institutes. Formal coursework and case conferences indoctrinate the analytic candidate into the body of knowledge that defines the epistemic community, namely, psychoanalytic theory. These courses focus exclusively on the baroque intricacies of the body of psychoanalytic knowledge—to the exclusion of other psychotherapeutic theories—and guarantee its stability and continuity among future practitioners. Thus, formal coursework provides the indoctrination that transforms psychoanalytic theory into psychoanalytic ideology for those who intend to practice the system of ideas.

Finally, there is the personal as well as professional experience of the training analysis. This process focuses on resocializing the candidate into a new symbolic universe of meanings, that is, a new reality. Although all of these processes are integrated in the institutes' programs, the critical experience is clearly that of resocializing the candidate through training analysis. It is in the training analysis that the critical function of having the candidate experience the integration of psychoanalytic theory and practice with individual personality structure and conceptions of reality occurs.

In the course of this experience, the candidate is forced to relinquish previously held images of reality by reinterpreting his/her past biography *en toto* and constructing a new personal and professional reality in terms of the psychoanalytic frame of reference acquired through indoctrination. Moreover, any reservations or disagreements the candidate has had about the validity of the theory and its interpretation during supervised clinical work are resurrected and examined as defense mechanisms and resistances that must be dispelled before the training analysis is complete. In other words, it is not merely the candidate's motivational equipment that must conform to institute standards, but also an acceptance of the body of knowledge around which the institute is organized. The institute in effect uses that body of knowledge and specialized techniques to guarantee the candidate's complete acceptance of the knowledge base of the epistemic community. Until there is complete acceptance of the specialized body of psychoanalytic knowledge, the training analysis cannot be completed; lack of acceptance of some aspect of the doctrine is viewed as reflecting potential psychological problems that would interfere with the proper practice of psychoanalysis.

This extended discussion of socialization of new members into the epistemic community of professional psychoanalysis has attempted to illustrate one important strategy that epistemic communities employ to delineate and maintain their boundaries and their autonomy; guarantee internal conformity and trust, as well as continuity and stability of the body of knowledge, and establish public stereotypes that will gain them external legitimacy. There is little doubt that the epistemic community of psychoanalytic professionals associated with analytic institutes has an unusually well-developed and institutionalized procedure for recruiting and socializing new members. Nevertheless, we believe that all epistemic communities faced with the task of ensuring the stability, continuity, and boundaries of their frame of reference against both intellectual-theoretical and technical encroachments develop similar mechanisms for transforming both the cognitive and the affective characteristics of potential members. That is, all epistemic communities attempt to protect their underlying frame of reference by resocialization processes that transform both the personal and the specialized, technically relevant identities of new members.

Contexts of an Epistemic Community: Medicine and Conceptions of Illness and Deviance

For this and the following illustrations, we have chosen certain aspects of the American medical profession. In this section, we explore some of the complexities involved in establishing and maintaining or expanding professional jurisdiction over sets of situations and problems. The right and responsibility to apply specialized knowledge, of course, requires the acceptance of the relevant specialized frame of reference, including knowledge criteria. It further implies a certain meshing of broad cultural models with those of the specialized epistemic community, resulting in its legitimacy and authority.

Medicine has obtained a virtually complete monopoly over the content and context of its work, as well as exclusive jurisdiction over determining what illness is and therefore how people must act in order to be treated as ill. Following Freidson (1970:224–227), a significant part of the meaning assigned to illness is that of deviance: specifically, the deviance that is imputed to the ailing person is seen as beyond deliberate, knowing choice and is essentially beyond one's control, that is, unmotivated. Furthermore, the symbolic apparatus surrounding the label of illness implies that what is wrong is determinable by rational (scientific) knowledge known to and manageable by a special class of professional practitioners (physicians). Hence, one should not judge a sick person, for that person is not responsible for himself/herself or for the undesirable condition. From this perspective, illness is a form of temporary, conditionally legitimated social deviance. It introduces the sick person to a process of social control that seals off the deviant sick person from nondeviants and pushes the former into professional treatment institutions geared to remove the deviance and restore a state of health.

Parsons (1964:258–291) has argued that the dominant American value of universalistically judged achievement means that an individual's deviance is more likely to be seen as a disturbance of capacity—that is, as illness—than is likely in other societies. Moreover, as a result of the emphasis on activity and achievement, the aspect of sickness most likely to be encouraged in the United States is demonstration of the motivation to cooperate in treatment and to return to functional health.

The result of this confluence of forces means that deviance is more likely to be considered a problem of health in the United States than a problem of law, religious purity, or political commitment. In this connection, it is possible to suggest that law deals with acts of imputed deviance for which the actor is held accountable and made to pay, while medicine deals with imputed deviance for which the actor is not believed responsible and therefore is treated rather than punished. While emphasis is placed on universalistically achieved performances and performance capacity, there coexists a general trend toward liberal humanitarian responses to deviance. This means that medicine, as the profession officially designated as competent to diagnose and treat illness, will have within its jurisdiction a large amount of deviant behavior; relatively more than in other societies in which medicine is far less extensively defined in relation to law or religion. Put differently, the increasing emphasis on the label of illness for serious deviance has been at the expense of the legal label of crime and the religious label of sin. It has narrowed the limits and weakened the jurisdiction of the traditional control institutions of religion and law. As Freidson so perceptively observes (1970:248):

> Quite apart from this legal position of religion in the United States, however, I would insist that over the past century it has, quite independently of its legal position, suffered significant decline in virtually all industrial countries as a source for viable definitions of deviance. So, too, though to a lesser degree, has law. Like Rieff, I believe that "the hospital is succeeding the church and the parliament as the archetypical institution of Western culture." The hospital is becoming such an archetypical institution largely through a process whereby human behavior is being reinterpreted. Disapproved behavior is more and more coming to be given the meaning of illness requiring treatment rather than of crime requiring punishment, victimization requiring compensation, or sin requiring patience and grace.

We return to the therapeutic conception of knowledge use in the last chapter, but in a much broader context than here.

Until quite recently, medicine occupied a relatively unimportant institutional position among agents of social control But with the growth of the natural sciences and their application to social affairs, behavior was increasingly seen as stemming from specific causes over which prayer, motivated choice, and will power had little control. Moreover, scientific discoveries led to the successful applied (medical) treatment of such problems. It was on the basis of this core of scientific developments that a vague halo of scientific authority emerged. This encouraged the wholesale extension of medical definitions of deviance—presumably grounded in the authority of science—into areas of behavior previously managed and administered by religion and law. Szasz (1964:44–45) has captured the flavor of this development in the following statement:

> Starting with such things as syphilis, tuberculosis, typhoid fever, and carcinomas and fractures we have created the class "illness." At first, this class was composed of only a few items, all of which shared the common feature of reference to a state of disordered structure or function of the human body as a physical-chemical machine. As time went on, additional items were added to this class. They were not added, however, because they were newly

discovered bodily disorders. The physician's attention has been deflected from this criterion and has become focused instead on disability and suffering as new criteria for selection. Thus, at first slowly, such things as hysteria, hypochondriasis, obsessive-compulsive neurosis, and depression were added to the category of illness. Then, with increasing zeal, physicians and especially psychiatrists began to call "illness" . . . anything and everything in which they could detect any sign of malfunctioning, based on no matter what norm. Hence, agoraphobia is illness because one should not be afraid of open spaces. Homosexuality is an illness because heterosexuality is the social norm. Divorce is illness because it signals failure of marriage.

What all of this attests to is a fundamental shift in societal frames of reference with respect to the general nature of and appropriate responses to social deviance. What has been called crime lunacy, degeneracy, sin, and even poverty in the past is now often defined as illness, and social policy has been moving toward adopting perspectives appropriate to the imputation of illness. The temper of the times has supported the claim that the proper management of deviance is treatment in the hands of responsible and technically expert (frequently medically supervised) professionals. In the absence of the labels of sin and crime, what is done to the deviant is held to be done for his/her own good, to help rather than to punish or incarcerate, even though the deviant may not desire the treatment which may constitute or involve severe deprivation. In fact, personal opinions about one's treatment are held to be irrelevant and are discounted because they are not based on professional expertise. As Wootton observes (1959:206): "Just because it is so much in keeping with the mental atmosphere of a scientifically-minded age, the medical treatment of social deviants has been a most powerful, perhaps even the most powerful, reinforcement of humanitarian impulses; for today the prestige of humane proposals is immensely enhanced if these are expressed in the idiom of medical science." Yet there is another sense in which this superficially humanitarian shift in the dominant frame of reference with respect to deviance masks profoundly disturbing tendencies to curb the individual right of equality and self-determination and to legitimate a profession's right to claim official jurisdiction over wide ranges of behavior.

The official jurisdiction that medicine has established now extends farther than its demonstrable capacity to cure. Moreover, success in gaining general acceptance of the use of illness to label any disapproved form of behavior carries with it the assumption that the behavior is properly managed only by physicians. Therefore, the fact that physicians are willing to deal with a particular form of behavior suggests implicitly that the behavior must be an illness. This means that medicine obtains official jurisdiction over areas despite the fact that it has neither knowledge of the etiology of the (presumed) condition nor a predictably successful method of treatment.

Alcoholism represents an excellent illustration of the expansion of medicine's jurisdiction over an increasing range of behavioral phenomena, despite the lack of scientific or moral basis for the legitimacy of its authority. Alcoholism used to be regarded as a sign of moral degeneracy and the drunkard was viewed as a sinner who sorely needed firm religious direction. With the advent of public inebriate statutes, what is now termed *alcoholism* became subject to the law and it was the legal profession that took responsibility for controlling the phenomenon through punishment meted out by the courts. Only in post-World War II America has excessive drinking become the province of the medical profession with the emerging popularity of "the disease concept of alcoholism" (Jellinek, 1960). Despite the fact that alcoholism does not fit the germ theory of disease model, despite the fact that it is impossible to distinguish symptoms from an identified disease entity, despite the fact that neither the etiology of alcoholism nor a predictable successful treatment modality are understood, public policy recognizes the authority of the medical profession to diagnose and treat alcoholism. This clearly has a certain humanitarian appeal: by labeling something like alcoholism an illness and declaring an appallingly drunk derelict to be sick—a condition over which one is presumed to have no control and therefore no responsibility—the humanitarian intention is to avoid moral condemnation. Thus, the illness is condemned rather than the person, but it is condemned nonetheless. In short, while ideally the person may not be judged, the disease certainly is judged and is clearly part of the person. And insofar as illness is defined as something bad or undesirable, medicine has become what Becker (1963: 147–163) calls a "moral entrepreneur." That is, medical practice leads to the creation of new rules defining deviance, and medical professionals seek to enforce those rules by attracting and treating the newly defined deviant. As Freidson observes (1970:255) "most of the activities of the active moral entrepreneurs of health are permeated by the tendency to see more illness everywhere around and to see the environment as being more dangerous to health than does the layman. . . . They are biased toward illness as such and toward creating secondary deviance—sick roles—where before there was but primary deviance." The recent United States government campaign against alcoholism represents an excellent illustration of this covert moral entrepreneuring on the part of professional medicine.

Although the growth in the use of illness as the principal label for social deviance is ultimately supported by the contemporary American emphasis on the value of health-as-generalized-performance-capacity, only the medical profession has an official mandate to discover, define, supervise, and treat illness. At this point, we can begin to employ some of the concepts introduced in this and preceding chapters to interpret the recent rise of medicine to preeminence among the institutions responsible for creating-defining, locating-diagnosing, maintaining-sustaining, and punishing-treating-rehabilitating deviance.

The growth of professional medical influence with respect to a wide range of phenomena that formerly were considered the province of legal or religious institutions is a consequence of the specialized frame of reference that characterizes modern medicine. First, medicine is committed to treating rather than merely defining and studying illness. It has a mission of active intervention guided by whatever it believes or suspects to be ill in the world. This mission is charged with a moral mandate to seek out illness—even to discover illnesses of which laymen may not be aware. In addition to this missionary stance of active intervention in anything that concerns health as well as illness, and almost anything can be interpreted as affecting the former, the medical frame of reference contains a professional decision rule that promotes constant expansion in cases of uncertainty. That is, the assumption that it is better to impute disease than to deny it, and risk overlooking an illness, becomes the basis for continuously expanding the domain under medical jurisdiction. For example, in the cases of alcoholism, mental illness, homosexuality, and divorce, to name but a few such conditions, there is great uncertainty as to the causal contribution of a number of variables. In such a situation the medical decision rule leads professionals to impute disease, just in case disease entities play a role in the condition. After all, the goal of medicine is to seek out, discover, and treat illness. This means, however, that medicine seeks to create new social meanings of illness where those meanings or interpretations were previously lacking. Once disease is imputed, medicine is given official mandate to manage the condition—even in the absence of any demonstrable capacity to cure or any desire on the part of the person displaying the condition to be rid of it.

This contrast between the medical and the legal decision rules (see also Scheff, 1966:105–127) and reality tests, and their differential consequences for professional expansion can be illustrated by considering illness and guilt, on the one hand, and health and innocence, on the other, as analogous. That is, both illness and guilt are conditions that are considered socially deviant and warrant further authoritative professional action. Conversely, health and legal innocence are considered normal and require no further action on the part of the professional or administrative authorities. The central thrust of the law is to look for and protect the innocent ("It is better to let a thousand guilty men go free than convict one innocent man"). In focusing on the presumption of innocence, the legal decision rule generates a tendency to protect individual rights and techniques of proper procedure. In contrast, the emphasis on illness built into the medical decision rule, the emphasis on diagnosing illness rather than health when doubt exists, generates a tendency to constantly expand the range of types of phenomena that come under medical jurisdiction and authority. Thus, the territory under medical authority and administrative jurisdiction is constantly expanding, at the expense of both legal and moral authorities. Moreover, because the scientific, biophysical status of medicine as an applied profession is confused with the moral and sociocultural status of the meaning of illness, its objectivity is unquestioned and few serious questions are raised about medicine's qualifications to enter basically moral domains.

Recently this expansion of the domain of medicine and the scope of situations with which the medical profession deals has come under attack. An emphasis on patients' rights, and the expectation of accountability for medical performance as it is expressed in the malpractice suit phenomenon, have also created a most complex domain of interaction between the medical and legal professions. Indeed, we might consider this as a battleground for what we have termed *situation control* by two major professions, even though this does not exhaust the issue of the revolt of clients against the authority of professions. Nowhere is the contrast in reality tests and decision rules more dramatically illustrated than in recent legal attempts to emphasize patient rights in cases of mental illness and hospitalization. The medical and legal modes of constructing these situations differ sharply. Indeed, the emphasis on patients' rights makes presumptions of rationality and responsibility that the medical perspective rarely shares.

More than most professions in post-modern society, medicine bases its claim for its privileges, prestige, and authority on the possession of specialized knowledge and a set of skills so complex that laymen cannot perform the necessary work safely and cannot even evaluate professional performance properly. Thus, the autonomy of the medical profession is based on the technical character of medical knowledge and work. Reliance on this point gives the physician legitimate exemption from supervision by nonmedical persons, even though the physician depends on the state for formally establishing and enforcing personal autonomy. The interplay between professional autonomy, a specialized professional frame of reference, and the significance and impact of the contemporary medical frame of reference that we have described above is well captured in the following quotation from Freidson (1970:330):

> Within its own institutions, protected by its organized autonomy, the profession has developed knowledge of its own and, by virtue of being a consulting profession, a capacity of its own to shape the behavior and experience of the layman independently of the lay community. In those institutions, the profession does not merely treat a biological state by biochemical or physical techniques: it also organizes the social identity of the layman into being a patient. Thus, in applying its knowledge, the profession cannot avoid making social as well as "purely medical" decisions about the people it deals with . . . when an occupation arises to serve some need or demand on the part of the lay community, and subsequently succeeds in becoming a profession, it gains the autonomy to become at least in part self-sustaining, equipped to turn back and shape, even create that need anew, defining, selecting, and organizing the way it is expressed in social life.

The quality of professional autonomy is likely to be quite different depending on whether it is based on the claim to role-embedded or implicit knowledge (*art*) or whether it is an explicitly knowledge-based profession with reference to scientific bodies of theory and data. To the extent that medicine was predominantly art, its professional structure and autonomy remained largely local and diffuse. However, as the domain of medical art is increasingly narrowed (it is unlikely ever to be eliminated) in favor of standardized, explicitly science-based technique, the mode of professional autonomy is likely to change. Explicitly knowledge-based performances can be checked without reference to particular community standards; they become amenable to the universalistic inquiry of both science and law. It is precisely this phenomenon, coupled with a remarkable knowledge transfer about basic health issues to the general public through education, that lies behind the more recent demand for medical accountability. Patients' rights movements as well as the rising frequency of malpractice suits are not only based on external forces but also on changes internal to the medical profession itself.

Contexts of an Epistemic Community: The Articulation of Interests in a Professional Association. The Case of the American Medical Association

In discussing the defining characteristics of epistemic communities and describing some of their strategies and unique attributes, we focused centrally on cognitive/cultural factors. That is, we gave central importance to those epistemic criteria for validating knowledge that represent the core of their various frames of reference. Using the professional medical community as an illustration, we focused on such things as the establishment and maintenance of professional autonomy for the epistemic community, the role of professional decision criteria in the medical frame of reference, and some of the most important boundary maintenance and exclusion mechanisms, as well as means of externally legitimating the specialized work of the professional group to the larger public in order to maintain the necessary degree of autonomy.

Throughout the illustrative discussion of organized medicine as an epistemic community with a specialized and well-differentiated frame of reference, we have avoided equating the professional medical (epistemic) community with its most visible, organized representative, that is, the American Medical Association. We have done this because although many professional epistemic communities, such as medicine, law, and the scientific disciplines, spawn professional associations to represent them, these associations are neither coterminus nor synonymous with the epistemic community itself. Thus, we suggest that in most cases the professional associations that represent the political and economic interests of specialized, technical epistemic work communities are rarely composed of the individuals actively or even partially engaged in the production, extension, or application of knowledge. Rather, we maintain that these associations are engaged in conventional pressure group politics geared toward the objective of defending, maintaining, and wherever possible enhancing the personal prestige, social status, and especially the professional autonomy of members of the epistemic community. In order to illustrate the subtle nature of the relationship between a specialized epistemic community with a distinctive frame of reference and the professional association that emerges to represent publicly and legitimate politically its autonomy, we will briefly consider the relationship of the American Medical Association to the larger community defined by the scientific medical frame of reference.

The professional associations that represent a large number of organized, specialized epistemic communities and that have played a critical role in making them the self-regulating, legally constituted bodies that they are today represent, in many respects, the modern counterpart of the medieval guilds (see Grant, 1942:303–336, on this point). The essential similarity between contemporary professional associations and the guilds is that the characteristic prerogatives of the guilds, as of the professional associations, were clothed with the authority of government, both national and local. Governments began to understand that the guilds were a central means of exercising their increasing responsibilities for the surveillance of industry. The guild system never explicitly took firm root on this side of the Atlantic in the frontier societies developing here, in which' most occupations were open to all who chose to enter. Therefore, it became necessary for such groups to fight many battles for recognition, legitimacy, and privileges that had previously been won in Britain and on the continent.

In the United States, the first medical licensure act was passed by the legislature of New Jersey in 1772 (Sigerist, 1935: 1057–1060). Analogously, the first attempts at legislative control of medicine in Canada were made as early as 1788, and the first medical act in Upper Canada was passed in 1795 (MacDermot, 1935). However, the American medical community did not organize a professional association until 1847. According to the literature endorsed by the A.M.A., it was "created to combat medical 'hucksters,' to root out commercialism, improve standards in medical schools, and raise the level of medical practice generally" (Fishbein, 1947). In contrast, Mills (1951) holds that the Association was a response to threatening competition and was formed to ward off this threat and win entrance to a broader area of economic pursuit. *A History of the A.M.A.* (Fishbein, 1947) proposes that the prosecution of quackery, nostrums, and secret medicine was the main reason that the Association was created.

At a more abstract level, all of these interpretations of the rise of the A.M.A. implicitly assume that protection and enhancement of both personal and professional status and autonomy were the central motivations underlying the organization of the A.M.A. Thus, the elimination of inferior competitors, the limited entrance to the field, and the improvement of quality of medical education and practice are all aimed at elevating the status and legitimating the autonomy of the medical profession in the person of the individual practitioner. And there was a definite functional need for the actions collectively taken by medical practitioners in 1847 when they formed their professional association: not only did hucksters and quacks exist, there was an expressed demand for their goods and services in the competitive market.

In contemporary and recent American society, physicians occupy one of the highest status levels of any professional category. It seems reasonable to presume that the physician's skills and knowledge are the chief considerations for what may be interpreted as an implicit indication of their relative value to the society. However, in earlier periods, the physician was assigned high status on the basis of values attached to the means of earning a living—namely, the ownership of physician's tools. This direct ownership of the means of production represented visible characteristics of a free entrepreneur, a role that was of particularly high value in nineteenth century America.

With the progressive advance of clinical medical science as well as medical industrialization and bureaucratization, which resulted in the hospital becoming the center of the production of the physician's goods (medical care services), the crucial value underlying this social status allocation involved specialized technical skill and knowledge, the critical attribute that resulted from this high social status of medicine was the institutionalization of professional autonomy. In view of the fact that the legal and other professions of equal intellectual capacity and technical skill are not all of the same status or autonomy as the medical profession, it is difficult to argue that social status and prestige, in and for itself, was the sole or basic motivation for the formation of the A.M.A. as a professional association. On the other hand, status problems impaired, if not precluded, the extent to which the professional (epistemic) community could attain and legitimate its professional autonomy and also impaired the widespread application of professional scientific medicine. It limited its capacity for situation control. Specifically, in the early 1800s the threat of unqualified medical practitioners was eminently real because they occupied the same status level as physicians and there was no legally constituted autonomy that protected scientific medicine from encroachments by other types of practitioners.

Medicine before 1840 was in a relatively anemic condition. Diagnostic methods were crude and the training of young physicians took place in private schools owned for profit by older, experienced physicians who desired a source of extra income and were thus assured of inexpensive assistants. The difference between medicine and quackery in reality was small and depended largely on idiosyncratic choice by the patients. With the importation of scientific medicine from Europe, however, this situation began to change. Improved medical techniques brought with them better diagnostic methods, including the necessity for intimate physical examinations and patient cooperation in answering sensitive, intimate questions. Physical examinations and the free, open discussion of matters intimately related to the human body violated powerful taboos. Moreover, the new medical knowledge enabled physicians to prescribe rigid regimens and, as they became more efficient, physicians saw increasing numbers of people. These aspects of the new kind of medical practice—and, especially, the invasion of privacy and physical manipulation—violated cultural taboos and were directly opposed to basic values and mores rigorously upheld in the expanding industrial society (Parsons, 1950:428–447 and 1964:258–291).

In order to overcome these potentials for social conflict, some novel kind of relationship between physician and patient had to be invented, legitimated, and established as conventional practice. In order for the physician to overcome various taboos and to remove any doubt that recommendations were being carried out, he needed a position in which acts of authority were unquestionable. In short, he had to establish his legitimacy and that of the new scientific medical order. It was the necessity for proving and justifying their professional status and autonomy that caused medical practitioners to create and maintain a professional association geared to protecting, maintaining, and enhancing their personal status prestige and professional autonomy. The members of society voluntarily submitted to the demands of the new professional association in the hopes of promoting better medical care; specifically, society awarded physicians the personal status prestige and professional autonomy they now possess.

This publicly legitimated high status proved extremely useful for physicians: in order for individuals occupying almost any status to accept the legitimacy of the physician's authority and autonomy, the physician required the highest possible status. The protection and maintenance of this status—and hence, the professional autonomy and authority that is dependent on it—was and is even now regarded as the responsibility of the professional association of scientific medicine, namely, the A.M.A. It has been the continuous task of the A.M.A. to prove that the behavior of society was justified, appropriate, and necessary, in the sense that the status and professional autonomy accorded organized medicine are responsible for more efficient and better medical care. In this view, any decline in either the status or the autonomy of medicine would result in a loss of professional authority as well as specialized technical expertise that would, in turn, affect the quality of care provided by physicians. Thus, the primary effort of the A.M.A. has been directed toward maintaining personal and professional status and autonomy for physicians. And the need for some instrument to fulfill this function represented the basic motivation for the medical profession to develop its professional association as it has and is today.

From one perspective, the direct beneficiaries of the activities of the A.M.A. are the public, in terms of the level of medical care and services provided, and the profession, in terms of the personal and professional status prestige and autonomy. "The A.M.A. is motivated both by obligations to the public and loyalty to its own members. The demands on it from these two points of view underlie all its activities suggest the possibility of conflict" (Hyde and Wolff, 1954: 938–1021). These potentially conflicting obligations and loyalties determine the objectives of the A.M.A. as well as explain some of the A.M.A.'s resistance to change.

The next paragraphs attempt to outline some of the more instrumental professional objectives of the A.M.A. based on the proposition that the central function of that professional association was and is to raise the quality of medical care. Our argument so far has been that this could only be done if the individual medical practitioner was granted the degree of social recognition, as expressed and manifested by social status and prestige, that would guarantee autonomy and authority in medical affairs.

It is possible to describe briefly the instrumental goals of the A.M.A. in terms of three immediate objectives:

1. Raising the quality of medical care;
2. Maintaining free determination of the conditions under which this care will be provided; and,
3. Educating the public.

Thus, we find the Association offering the following description of its purposes:

Its purposes are to promote the science and art of medicine; to organize the medical profession and safeguard its interests; to elevate the standards of medical education and practice; to bring about the enactment of uniform legislation for the public welfare, and to protect the public health.

(A.M.A. News, 1947:1)

Although it might seem that the free determination of the conditions of practice and public education would be more immediate means toward the desired objectives, the repeated insistence by the A.M.A. on its interest and role in raising the quality of medical care is indicative of a number of different group references. The interests of the professional membership clearly are more directly protected by achieving the latter two objectives. However, for the individual physician and, incidentally, for the public at large, the improvement of medical techniques and the dissemination of new facts are of more direct interest than such factors as method of payment and involved legal and economic arguments relating to public health economics. The activities of the A.M.A. relating to medical practice per se are probably best known to the membership and least well realized by the public. The A.M.A. has been especially active in controlling the quality of products that are involved in the practice of medicine and in raising standards of medical education, hospital administration, and medical research methodologies.

Because of the physician's interest in the quality of medical care, other nonclinical aspects of medical practice are inappropriately related to the heretofore elusive criterion of quality and then inappropriately evaluated in terms of some imaginary, ambiguous criterion of proven higher quality of medical care.

> The criterion upon which approval is based, is the requirement that a high quality of medical care and medical service be made available without interfering with the physician-patient relationship. The A.M.A. would probably support any suggested alteration in furnishing medical care to the Nation which would in reality guarantee a higher quality of service.
>
> *(Bortz, 1947:565-570)*

This insistence on proven higher quality before any "alteration in furnishing medical care" leaves little room for experimentation and results in highly conservative responses to political issues, since past experiences have indicated a relatively (although not uniformly distributed) high quality of medical care in contemporary America. The professional association of organized medicine attempts to justify its conservation by pointing to past achievements, and the political biography of the A.M.A. is permeated by historically oriented perspectives. In fact, the A.M.A. only endorses new features of medical care, medicine, or public health if they are completely compatible with an established historical tradition.

The free determination of the conditions under which medical care is provided, in terms of methods of practice and the free choice of physicians, is the area where the A.M.A.'s activities acquire a more explicitly political flavor. The political overtones of these issues insure that they are widely publicized by the A.M.A., in order to reach large segments of the population with what is regarded as an important aspect of medical education. However, it is important to emphasize that all of the A.M.A.'s activities are considered equally important to the Association itself and none of them could exist independently of the others. That is, in order to insure the professional authority and autonomy of physicians, their ability to practice medicine must be unique and above reproach, and this in turn requires the high, personal, status prestige that medical professionals are accorded —at least in the view of the professional association of scientific medicine. Thus, the A.M.A. has spawned its own Councils on Medical Education, Pharmacy and Chemistry, and on Medical Services, as well as Bureaus of Investigation, Legal Medicine and Legislation, Medical Economics, etcetera.

One essential way of effectively defining, structuring, and controlling the context of medical practice is to have all legal authority of the state delegated to a professional association, the A.M.A., insofar as medical practice is concerned. American custom has left the conduct of medical practice in the hands of the medical professional association; this was necessary once society accorded medicine the privilege that only another physician could judge the validity and competence of a colleague's activities. Although the licensing of physicians is now a state function, the A.M.A. is still able to control such licensing through control over the members of the Licensing Boards. Moreover, when legal matters in the licensing of a physician are considered—such as citizenship, criminal background, etcetera—one finds that the professional association has in effect delegated these powers back to the state and, in so doing, strengthens its position as the ultimate arbiter of medical affairs.

The A.M.A. insists that its advocacy of free determination of methods of practice and of free choice of physicians is inseparably related to the preservation of the authority and autonomy of medical practitioners and, consequently, to the quality of medical care provided. Free determination of methods of practice, which includes method of payment, essentially means "no state interference, except for rigid state licensing systems" (Mills, 1951). Again, the perspective underlying this stance is simply that any external or extraneous influence involves a threat to the profession's autonomy and authority and, hence, to the quality of medical care the public receives. In general, the A.M.A.'s defense against these threats takes the form of describing the grave dangers to the quality of medical care that inhere in any other form of payment than the fee-for-service method payment, which is often treated as if it were solely responsible for any successes in improving the nation's health. The Association's communication media continuously claim that any methods of practice other than fee-for-service would allow, if not force, accounting and controls by others and this would immediately effect the way a physician renders service such that it would be detrimental to quality. The real objection underlying this position taken by the professional association is that accounting controls presuppose a higher authority than the medical association itself. In short, the A.M.A. has been largely successful in its attempt to obtain legal backing for its views and positions with respect to medical affairs and the Association appears to have a number of powerful mechanisms through which it can effectively control national and state health legislation relevant to the public's health.

The broad scope of this desired sphere of influence forces the A.M.A. to involve itself in major public education campaigns that repeatedly assert its value and the value of its members to the public. "Today's Health" and scores of pamphlets, newsletters, and exhibits are used by the A.M.A. to obtain public understanding of the medical professions's sociopolitical position, objectives, and practices. The Association assumes that increasing understanding will result in reinforcing the legitimacy and autonomy of the medical profession. In view of the fact that the A.M.A. has lost few major legislative battles, the Association's methods and media of public education and opinion formation appear to be extremely efficient and effective in achieving the ultimate objectives of the professional association. In short, there is no unambiguous reliable evidence that the personal status prestige or the collective professional autonomy of physicians have been reduced to any appreciable degree in recent times. Moreover, the technical quality of specialized medical care seems to be constantly improving, albeit in rather an unevenly distributed fashion. Whether or not any past or recent improvements are actually related to the professional association's preservation, maintenance, and enhancement of professional autonomy and authority as well as the personal prestige of physicians is, in a very real way, anybody's guess.

The professional association self-confidently assumes that the present quality of medical care can be directly traced to the autonomy and authority of physicians, which the Association argues are upheld by the high personal status prestige of the individual practitioner. In view of organized medicine's very positive evaluation of the technical quality of medical care provided in contemporary America, the professional association created to protect the interests of the organized medical community can be said to have functioned with remarkable success and effectiveness. From a different point of view, in terms of protecting and guaranteeing the continuity and stability of the scientific frame of reference that defines the work of the medical community, the A.M.A. must be given considerable credit for its success in resisting changes.

The A.M.A.'s membership is entirely composed of physicians, although they in no way represent a homogeneous group of professionals with a certain communality of personality, as the Association frequently assumes. The fact that about one-third of American physicians do not belong to the A.M.A. does not indicate that the Association or the larger epistemic community is falling apart. It simply means that the professional association does not control all activities in which physicians participate. Four groups of professionals do not have any need to be members of the professional organization in order to be physicians: members of university faculties; government officials who are not dependent on A.M.A. controlled aspects of medical practice; physicians in private research or administrative functions; and retired physicians. But aside from these four groups of practitioners, the sanctions for not belonging to the professional association are extremely strong and are rigorously enforced. Refusal to enter, denial of admission, or expulsion from the Association is likely to incapacitate the individual practitioner in the performance of normal roles. Specifically, nonmembership usually means that the physician will be denied the facilities of certain hospitals and that other physicians will frequently be extremely reluctant to enter into consultation relationships. Moreover, insurance agencies will demand higher premiums for malpractice policy—if they will insure the physician at all.

Procedurally, the American Medical Association has developed an implicit norm that specifies that all organizational policies and decisions must be accepted unanimously by members. This norm is based on the belief that the members all have the same social and professional objectives and values by virtue of their identical medical school training and selection procedures. Thus, the professional association assumes a certain communality of personality that, in turn, generates the belief that all physicians must necessarily have the same goals and employ the same means to these goals. Nevertheless, the professional association faces a staggering problem in the task of molding, coordinating, and modifying the opinions of the large number of members who practice under widely divergent contexts and conditions throughout the nation. The most important mechanism for overcoming problems of geographical distance and professional differences are systems of communication developed by the Association: for example, publication of the *Journal of the American Medical Association* and the *A.M.A. Newsletter*, which are the instruments for molding medical opinion on a week-to-week basis.

In spite of a heavy volume of professional literature, most members of the medical epistemic community are not very familiar with either the objectives, the structure, or the specific actions of their professional association. The individual physician is largely politically ignorant and apathetic. This is hardly surprising since the physician's training and, prior to that, acceptance to medical school have narrowly oriented the doctor to illness and the sick patient needing treatment. Physicians are generally unprepared to accept collective responsibilities. This apolitical, apathetic ignorance is one of the most pressing problems of the A.M.A. in terms of potentially influencing popular attitudes about particular actions of the medical profession. The strong primary ties that emerge in the patient-physician relationship make it ideal as a context for explaining the A.M.A.'s rationale for a particular policy and thus initiating a process of molding public opinion. Yet the A.M.A. is seriously impeded in its efforts to use this primary group relationship as an effective method of influencing popular attitudes on medically relevant issues by the lack of enthusiasm that physicians seem to have for using this context as a forum.

Judging from the lack of knowledge about the Association among physicians, and judging from the amount of misinterpretation about the A.M.A. which constantly appears in public print, it seems evident that you do not sufficiently exercise intelligent interpretation of the Association to your members at home.

(A.M.A., 1953:41)

In order to rectify and prevent further misrepresentations and ignorance of official organizational doctrines and objectives, practicing physicians and medical students receive, without provocation, such materials as "It's Your A.M.A.," "Economics and the Ethics of Medicine," "New Forms of Medical Practice," "An Introduction to Medical Economics," etcetera. In short, the A.M.A.'s task in obtaining uniformity of normative behavior and commitments to the organization is a formidable one; mass education of the membership seems to be the central way in which the A.M.A. attempts to grapple with this task:

> You are the A.M.A., you share in its prestige and its strength. You must also bear a share of the criticism it receives. Your association asks from you the loyalty to refrain from attacking it without knowledge; the loyalty of working from within to effect change, rather than the throwing of rocks from without; the kind of loyalty that will make a member accept criticism of the A.M.A. as criticism of himself, and, if the criticism is unjustified, defend his association, or if it is justified, take steps to correct the situation.
>
> *(A.M.A., 1953:41)*

Finally, how does the professional association that is formed to handle the political issues of a specialized epistemic community attract members of that community to serve as the leadership elite of the professional association? Chapter 3, Article 1, Section 2 of "Principles of Medical Ethics" (A.M.A., 1953) provides a guide to the answer:

> In order that the dignity and honor of the medical profession may be upheld, its standards upheld, its spheres of usefulness extended, the advancement of medical societies promoted, the physician should associate with the medical societies and contribute freely of his time, energy and means, in order that these societies may represent the ideals of the profession.

The physician that is able to "contribute freely of his time, energy and means" to the professional association promulgated by the epistemic community is likely to be a person who has achieved professional recognition, financial independence, and social prominence. One must be able to leave a practice frequently to attend meetings, while other activities connected with the office will tend to restrict this ability to practice medicine actively. Through the years, the A.M.A. has been dominated by the older members of the profession, and it is not surprising that these physicians tend to be the more conservative ones (Garceau, 1941:81).

These physicians tend to be specialists, rather than general or family practitioners. "All but one of the present (1954) members of the Board of Trustees and all A.M.A. presidents, vice-presidents and speakers since 1947 have been specialists" (Hyde and Wolff, 1954: 938-1021). The distribution of physicians being what it is, the specialists tend to come from urban areas. Specialization and urbanization are related to financial independence, since specialists average more annual income than general or family practitioners. Furthermore, professional recognition is best achieved in large complex hospital settings and, hence, is facilitated by urban as opposed to rural practices. These considerations also undoubtedly lie behind Garceau's (1941:50) observation that members of the A.M.A. House of Delegates who served long terms are primarily from large population centers whereas shorter term delegates are more likely to come from rural areas. From the perspective of the overall political stance of the A.M.A., the specialized, industrialized method of medical practice and generally more liberal urban point of view tend to make specialists somewhat less conservative, on the average, but the prerequisites of age and affluence apparently neutralize that tendency so that the balance swings back in favor of a conservative resistance to political-medical change.

The motivations underlying the desire for leadership in a professional association elite are legion, but it is probable that feelings of responsibility for public service represent, at least in part, one relatively powerful impetus. Since the administrative skills necessary to leadership can become highly developed in modern hospitals or clinics, specialists, who are far more likely to practice in these settings, are almost automatically selected. Moreover, the public and political character of the A.M.A. is such that some previous experience in dealing with groups is desirable in order to overcome personality conflicts. The private general practitioner usually does not deal with groups and probably is unaware of the range of problems that are involved in collective negotiations, as his/her interests lie in treating the sick patient as an individual. In fact, it does not seem exaggerated to suggest that the personality characteristics demanded by the functions and roles of leadership in professional associations are generally incompatible with the personality characteristics that are created, selected, and reinforced by the free practice of medicine. This may partially explain why rural members of the House of Delegates tend to stay for generally such short terms of office.

Naturally, personal advantage may also be a strong motivation for belonging to the membership elite of the professional association. "Politics has saved many tolerable but undistinguished doctors from the economic disadvantage of oblivion" (Garceau, 1941:64). Such persons would probably enjoy the numerous social contacts, traveling, recreation, publicity, and professional recognition that are the rewards of office. Moreover, physicians having this motivation would be more likely to master parliamentary rules than colleagues in office for only a short period of time. Since personal advantage can only be realized by adequate performance of the public responsibilities associated with an office, it seems certain that the reward schedule is structured in a way that will tend to induce responsible behavior in office holders in enough instances that the organization can function properly in implementing its objectives.

The objective of this illustration has been to describe the American Medical Association as an example of a highly organized professional association formed to protect, maintain, and advance the interests of the scientific medical epistemic community. Our purpose was to demonstrate that the professional association developed by a specialized epistemic community is clearly different from the epistemic community itself. Therefore, in describing the A.M.A., we have avoided all mention of the core characteristics of medicine as an epistemic community: that is, we have specifically avoided any reference to the specialized medical frame of reference, to medical decision criteria, to both scientific and professional medical reality tests, and to specialized technical training and socialization procedures for inducing new members into the medical frame of reference. Our central point has been to demonstrate that professional associations are a structural arrangement by which epistemic communities deal with the external problems of guaranteeing public support and legitimacy for their activities and that these external, political instruments used by epistemic communities are quite separate from the communities themselves—even though the lay public tends to equate the visible professional association with the epistemic community it publicly represents in the political arena. Yet, as the preceding discussion has shown, professional associations are centrally concerned with political and social resources and legitimacy, whereas epistemic communities are primarily oriented to some aspect of the production, application, and codification of a body of knowledge.

The Transformation of a Professional Epistemic Community: The Emergence of Family Medicine and Family Health Care

American medicine appears to be in the midst of a rapid swing of the pendulum from a system of medical education firmly rooted in biomedical science, and oriented toward a high degree of specialization in practice, toward one that consciously attempts to give greater emphasis to a more humanistic attitude, to the social and behavioral sciences, and to more generally (family) types of practice. In order to understand the reasons for suggesting that this rather sudden swing toward family medicine, community medicine, and primary care represents a profound transformation in the ideological and professional context of the medical epistemic community, it is necessary to look at the current situation in medical care in the United States, specifically as it concerns manpower, costs, and content.

Manpower and Access to Medical Care. The problem of medical manpower has at least three central aspects: numbers, geographic distribution, and distribution of available personnel among functional categories within the various health professions. Each of these aspects relates to the others. If every citizen is to have ready access to health care, suitably trained purveyors of primary health and medical care must be distributed in sufficient numbers throughout the country to meet the health needs of local population groups. The problem concerning the number of health personnel has been debated endlessly and still is not settled. Until recently, the Federal government consistently stated that America needed a greatly increased output of physicians. Within the past decade, however, government analysts appear to believe that we are approaching adequate numbers. But most authoritative observers (exempli gratia, Magraw, 1966) agree that the most serious problem in the American medical care system is the shortage of providers of primary care and an excess of specialists—in other words, maldistribution within the profession.

This crisis in the medical epistemic community represents a consequence of reforms in medical education brought about by the Flexner Report (Flexner, 1910). The essence of those reforms involved insistence on placing medical education on a scientific basis. This led to a closing of many low quality, nonscientific medical schools, to a reduction in the total output of licensed physicians, and to a shift of clinical instruction into teaching hospitals under the professional control of fulltime clinical faculty, who were generally selected for their research accomplishments. These faculty members inevitably were centrally preoccupied with the care and study of life-threatening disease episodes that brought patients into the hospital wards.

This trend was reinforced in the period after World War II by the massive infusion of federal funds into American medical schools and teaching hospitals for biomedical research and research training. By far the greatest amount of these funds to support new developments in the schools was not given directly to improve medical education—although it may have unintentionally had that consequence—but rather to cure or eliminate certain dread diseases, the choice of which depended to a large extent on the vagaries of Congressional whim or on the strength of various lobbies of influential citizens. In short, one of the unintentional effects of the reforms generated by the Flexner Report and the tremendous growth of biomedical research funding was to isolate medical education increasingly from the basic health care needs of the people and to concentrate it in medical centers delivering mainly tertiary care. Increasingly, clinical faculties in medical schools came to be composed mainly of scientifically trained specialists. Students were seldom exposed to the daily practice of medicine in the community or to those physicians whose role it is to deliver the primary and secondary care that constitutes the bulk of all medical practice.

The result of these developments was a steady decline in the number of general practitioners and a continuing rise in their average age, since few students entered this unfamiliar form of practice. At the same time, the absolute number and proportion of specialists steadily increased (Magraw, 1966). The extent of this trend may be highlighted by comparing the situation in the United States with that in Great Britain, where medical manpower has been controlled by the National Health Service since World War II. Specifically, Great Britain has a ratio of nearly three primary-care physicians to every specialist, whereas there are approximately four specialists for each general practitioner in the United States.

The deficiency in primary-care medicine in the United States has become increasingly visible in both urban hospital emergency rooms and in rural areas totally without physicians. Responses to this situation have been dramatic, albeit somewhat chaotic, during recent years. In 1966, the American Medical Association's Council on Medical Education published in the Mills Report, which recommended a shift in graduate medical education away from specialization toward the training of primary-care physicians (The Graduate Education of Physicians: The Report of the Citizen's Commission on Graduate Medical Education, 1966). This was followed by the Willard Report, which recommended the establishment of family practice residencies in line with the change in name of the American Academy of General Practice to the American Academy of Family Practice (Willard, 1966). These proposals or shifts in the trends of post graduate education of physicians received a further impetus from federal legislation that authorized funds for the establishment of departments of family practice in medical schools and for the development of family practice residencies. And although the actual funding of these educational departments and programs has been and remains uncertain, many institutions have enthusiastically moved into this professional area, lured on by the promise of government support. The extremely rapid growth of family practice residency training programs is attested to by the fact that between 1963–64 and 1973–74 forty-three family practice residencies were started, enrolling a total of 798 residents (Crowley and Leymaster, 1973).

The other response to the manpower crisis in medical care has been to increase the number and size of medical schools so as to achieve a substantial increment in the number of physicians. Ultimately, this should have two consequences: a wider distribution of physicians as opportunities of lucrative metropolitan practices decline and a lower income for all physicians. Another more rapid and undoubtedly less expensive approach to solving the primary-care manpower crisis involves the training of other health professionals—nurse practitioners, nurse associates, and physicians' assistants. This training would make it possible for these practitioners to include certain aspects of health supervision, home visiting, assessment of acute illness, and triage of patients in their professional activities. The object of this approach is to extend the physician's ability to give comprehensive care to a larger group of patients and to concentrate more exclusively on aspects of diagnosis and treatment in which long and expensive medical training can best be utilized. What is particularly important to note, for our purposes, is that many of these reforms aimed at dealing with the manpower crisis in the American medical care system have the consequence of reducing the autonomy of the professional medical epistemic community. Specifically, the introduction of massive federal funding along specified guidelines, the increase in the number and size of medical schools associated with it, and the training of other health professionals and paraprofessionals all have the impact of reducing the exclusive monopoly that the medical epistemic community has over the definition of health, illness, and treatment.

The Cost Crisis in American Medical Care.   For the past decade the cost of medical care in America has risen far more rapidly than that of any other major item in the cost of living (Rogers, 1973). Again, comparison of per capita costs of medical care under the British National Health Scheme and the American nonsystem of medical care shows striking differences for two highly developed countries with at least roughly comparable standards of living and health. With nearly twice as many physicians in proportion to population, health care costs per capita in the United States are over 3.5 times as high as in the United Kingdom. Moreover, British figures for infant mortality, which is usually taken as a relatively reliable index of health care, are considerably better than those in the United States.

Although there are numerous causes for the skyrocketing costs of health and medical care in the United States, certain factors seem particularly crucial. First, there is the overabundance of specialists; secondly, American voluntary insurance programs—Blue Cross and Blue Shield—encourage hospitalization rather than less expensive ambulatory care; thirdly, administrative costs within the hospital are raised by the necessity for itemized charges for every test procedure and medication; fourthly, inadequate primary care for urban and rural poor allows disease to become severe before patients are brought in for definitive treatment; finally, there are the activist, interventionist medical philosophy and disease criteria that dominate American medicine. Some of the basic dilemmas and difficulties involved in technical, specialized knowledge application in relation to costs and effectiveness can be illustrated by considering the concept of the annual physical or work-up.

The annual physical, or the more elaborate multiphasic health screen—a routine exam complimented by a battery of diagnostic tests—has been introduced as the cornerstone of virtually every new public or private program designed to maintain the nation's health. At the same time, the National Health and Medical Research Council of Australia noted the "multiphasic health screening procedures appear to be of little value in medical practice at present, particularly in respect to individuals who are apparently well." This sentiment appears to be shared by most of the rest of the world for, aside from a few isolated experimental projects, the multiphasic health screen has been a uniquely American phenomenon. In fact, it appears to have succeeded in achieving the unfortunate double-barrelled combination of low efficiency at very high costs. It is instructive to examine the historical origins of the popularity of the annual physical and associated work-up in America.

The concept of the annual physical examination was first promoted in this country around 1924. At that time, there was little else that could be offered to patients since, by today's standards, physicians were relatively impotent. Even though patients were urged to contact doctors regularly in order to enable early detection of disease, the fact of the matter was that there was precious little that could be done about diseases even if they were detected in their early stages. The technological advances in medical treatment that occurred during the ensuing decades clearly benefited the ill, and the promise was held out that these advances could also be applied to the healthy. The public was assured that most diseases could be cured if detected early enough. By the mid-1950s, the multiphasic health screens had been developed and become entrenched. The idea of receiving an expensive clean bill of health became transformed by legislative and judicial fiat into well-intentioned programs and slogans that made health, like liberty, an inalienable right.

By the 1960s, American medicine was undergoing serious criticism for its emphasis on the diagnosis and treatment of disease. Preventive medicine, public health, and individual superhealth became the vociferous demands. Politicians courageously responded to these demands with ringing endorsements of health. As the concept of the prepaid health maintenance organization (H. M. O.) gained momentum, the periodic annual multiphasic screening of the healthy became the cornerstone of the health care of an increasingly vocal force of health care consumers. Yet an increasing number of critics maintain that except for the detection of hypertension and a few other abnormalities, current data indicate that the periodic health exam, even when complimented by multiphasic profiles, is a costly and wasteful indulgence. The reason for this is that most diseases can be detected only after symptoms appear. Furthermore with the exception of hypertension, there is a paucity of convincing evidence that treatment of diseases before the onset of symptoms offers any long-term advantage over treatment that is initiated after symptoms arise. Although most of these critics acknowledge the value in encouraging people to have a single, limited physical examination as a means of establishing contact with the physician, most argue that further repetitive annual exams of healthy individuals is profitable only for the medical profession. In short, a marked tendency to use expensive broadsides of tests as well as other forms of overtreatment accounts for a considerable part of the high cost of hospitalization in America, not to mention a considerable volume of iatrogenic disease.

These and other cost factors are further complicated by the fact that third-party payers—either voluntary insurance companies or government agencies—now pay most of the bills although these costs eventually land on the consumer in the form of higher insurance premiums and/or increased taxes. The immediate result has been a spate of new regulations to control accelerating costs of medical care in America. Since medical care is now very close to being the single, largest economic activity in America, it is inconceivable that the present crises in medical education, manpower distribution, and health care delivery will be allowed to continue for long when the economic price tag is so high.

    The Shifting Content and Focus of Medical Care.   The traditional tasks of the physician have been the saving of life, the alleviation of suffering, and the minimization of disability. The prime concern of the physician was with the struggle against illnesses in the individual patient. The traditional responsibilities of the physician for saving lives persist in the early recognition and prompt treatment of medical emergencies. However, these episodes are rare in comparison with the more subtle and infinitely more complex problems of adjustment that give rise to disease in the daily lives of the patients seen by most primary-care physicians. Yet medicine continues to be primarily illness focused. This emphasis made sense when the major concern of most physicians was with the treatment of prevention of diseases having a single external cause. With increasing advances in medical science, however, it has been discovered that most disease—even specific infections or particular vitamin deficiencies—actually have multiple causes. For this reason, various critiques of American medicine have argued for the need to emphasize not the disease but the patient in his/her own setting (for example, J. B. Richmond, 1969, and R. N. Magraw, 1966). Under this reformulation of the basic medical orientation, the process of diagnosing disease gives way to evaluating the patient. This reformulation of the nature of medical practice emphasizes the personal bond between physician and the patient in the natural environment—most especially in the context of the family—and emphasizes continuity of care over the entire life span and across all family members.

    All of these deficiencies in American primary care have contributed to the transformation in the professional medical epistemic community that began less than two decades ago. In a technical sense, primary care is delivered by any physician who is willing to practice in a community and to see anyone who seeks him/her out for medical advice as the first physician of contact. In the United States until the last decade, primary care has been given by a dwindling number of general practitioners and by pediatricians and internists, in solo or group practice, in neighborhood health centers, and in the emergency rooms of urban hospitals. The recently established medical specialty of *family practice* is an attempt to provide primary care by physicians specifically trained for their job as specialists. Thus, family practice represents an established medical specialty embodying one model of delivering primary health care services. In theory at least, family practice means comprehensive continuing care provided to all age groups and all family members by a personal physician. At present, family practice often represents primary care that is simply a mix of medicine, pediatrics, psychiatry, office gynecology, and minor surgery and is in danger of being crystallized into a rigid model of super general practice by certifying boards. This danger is a result of a largely well-intentioned desire to set standards for the large number of new resident-trainees emerging from newly created federally funded residency programs before a real academic discipline of family medicine has been fully developed.

The end point of the contemporary transformation of the medical epistemic community will involve a combined discipline/specialty of family medicine that is concerned with the study of the family as an ecological social unit—a system in which the symptoms in one member may actually reflect disturbances in the life of another (Ransom and Vandervoort, 1973). In short, the crisis in primary care that has confronted the professional medical epistemic community has led to the emergence of an embryonic discipline of family medicine concerned with the relationship of life in small groups to health, illness, and care. Its focus, ultimately, will undoubtedly be on the ecology of relations among individuals in families and between families and their surrounding environments. (It is highly appropriate that such a radical definition of the core task of the medical epistemic community employs an equally radical conception of its subject matter, the family, as any group of intimates with a history and a future.) Put differently, the emerging discipline/subspecialty of family medicine aims at understanding and changing health problems that cannot be managed successfully by dealing exclusively with the individual and his/her illness abstracted from the pattern of recurrent interpersonal situations that shape and transform a human life.

Taken together, these developments represent a transformation of the professional medical epistemic community in response to widespread public perceptions of a crisis in primary health care. We have suggested that the response to this crisis has been a shift away from disease-oriented treatment of individual patients toward family and community medicine with its stress on preventive health care practices and health maintenance within specified social contexts. Another perhaps even more profound transformation in the professional medical epistemic community has involved the restructuring of medical ethics as a result of new technological controls over life and death. In order to understand the manner in which technological advances have rendered obsolete and untenable the central premise previously underlying the activities of the medical community, it is necessary to describe the classical ethical promise of scientific medicine.

Parsons, Fox, and Lidz (1972) have argued that the structural core of traditional medical ethics involved an absolutizing of the value of preserving life. That is, both the life of the individual patient and the physician's obligation to protect or save the patient's life have been taken as divinely given. This meant that physicians could then take the obligation to preserve and enhance patients' lives as an absolute prescription or commandment having no limitations whatever. Only insufficiencies of the physician's technical resources could limit efforts to combat the patient's death.

This ethical position had a number of extremely important consequences for the growth in the autonomy and authority of the medical epistemic community. First, it provided very strong ethical motivation to increase the specialized, technical capacities of the physician in order to better implement the ethical obligation to save lives under an increasingly broad range of difficult circumstances. Second, it provided an ethical basis for the autonomy of the professional medical epistemic community in that it required physicians to pursue the saving of life at almost any cost; that is, by subordinating almost all other value considerations. This meant that the nearly absolute commitment to preserve life strongly insulated medical ethics from any ethical system that did not place equivalent emphasis on the value of preserving life, thereby providing an authoritative grounding for the autonomy of medical ethics and of the professional epistemic community that represented them.

This classical ethical premise emphasizing the dignity and importance of divinely given human life, associated with the development of professional medicine, contained a number of difficulties that have recently begun to overwhelm it. One central dilemma is that this ethical pattern allows little room for positive definitions of the significance and meaning of death. Death came to be seen as a medical defeat, either for the physician personally or for the "state of the art." This placed physicians under great stress in circumstances under which they were bound to lose many patients. It led to the frequently documented inability of physicians to deal openly and frankly with dying patients. It also has produced situations in which treatment has been given more as a ritual commitment to the value of fighting death than out of any rational expectation that it will help the patient. It frequently and increasingly meant that the most intensive and sophisticated treatment technologies were applied to aging patients dying from degenerative diseases that could not be alleviated.

Under these circumstances, treatment often seemed to contribute more to unwanted suffering and pain in dying patients than to preservation of the gifts of life. Moreover, the traditional ethical focus on the obligation to defeat death through mobilizing intensive commitments and enormous quantities of medical resources had disastrous effects on the allocation of medical resources. The American medical community is somewhat unique in providing the most intensive, most technically elaborate and expensive treatments for patients who are already engaged in losing struggles with death or debilitating diseases. In these kinds of circumstances, there has been the tendency to attempt to demonstrate that everything possible has been done to aid the patient, regardless of the social and/or emotional costs to the patient, relatives, or society. Finally, the stress on medical heroics derives from the traditional ethical premise underlying the medical epistemic community. It diverted both attention and resources from public health measures that would have preserved far more lives at considerably less cost.

A variety of technological and scientific developments has undermined both the authority and autonomy of traditional medical ethics and has been responsible for the emergence of a new, more relativized medical ethic of responsibility. One crucial set of developments has concerned the understanding and definition of death as the end of life. In this respect, the recent trend has been toward redefining death so that brain function, rather than breathing or heart function, serves as the criterion of life. Moreover, recently developed resuscitation and life-supporting techniques have provided physicians with far larger numbers of patients on the border between life and death. Thus, physicians now do not merely save lives, but actually bring people back to life. In addition, increasing numbers of hospital patients lack, at least temporarily, the capacity to live but, rather, are kept alive artificially.

The clear-cut boundary between life and death has also become more ambiguous with respect to the origins of life in conception and birth. The ethical issues generated by birth control methods, and, especially, by the abortion of viable fetuses have become very difficult to resolve within traditional medical ethics. Most attention has been given to the question of when, in the development from conception to birth, the fetus obtains the gift of life that makes it sufficiently autonomous to be entitled to legal and moral protection. But perhaps the technical, therapeutic innovation that most dramatically exemplifies the ethical transformation occurring in modern medicine involves human organ transplants. This is because a central aspect of transplantation is what Parsons, Fox, and Lidz (1972) refer to as "the theme of the gift." Transplantation entails the most literal gift of life that a person can proffer or receive. The donor contributes a vital organ to a terminally ill, dying recipient in order to save and maintain that person's life. Moreover, live organ transplantation often involves both the donor and the medical team in an unprecedented act that violates traditional medical ethics: the infliction of deliberate injury on a healthy person in order to help another, who is suffering from a fatal disease. In this sense, it compromises the traditional medical ethical requirement "to do no harm."

All of these and other pressures on the traditional form of professional medical ethics have become so powerful that they have forced the medical epistemic community to undertake a very general reexamination of its morals. Thus, technological advances, as well as public pressure on the part of the clients or consumers of medical care, have challenged the autonomy and authority of traditional medical ethics to the point where a new, more relativized medical ethic is emerging that no longer treats the priority on protecting life as an absolute. That is, the physician is no longer under an absolute commandment to preserve life, but may make a relatively free judgment—after consultation with colleagues, the patient, and the patient's family—about whether treatment should be directed toward the preservation of life at all costs or whether other ends should be given priority.

This emerging ethical system in the medical epistemic community acknowledges explicitly the conditions under which direct struggle with death is fruitless or even counter-productive in terms of other values. It provides a sounder basis for confronting the problems of meaning that recent technological developments available to the medical epistemic community have been generating. This relativized medical ethic also provides a sounder basis for elaborating the positive or consummatory meanings of death. In consequence, the impending or inevitable death of a patient need not be taken as a personal defeat or a defeat of treatment by physicians. Most importantly, physicians can now attempt to facilitate the patient's death in a manner that supports the patient's dignity and ability to put affairs in order so as to meliorate the trying conditions of death for both the patient and the family.

This new, relativized ethical system underlying the activities of the medical epistemic community imposes some very serious new responsibilities on the physician. Physicians no longer have the emotional support of an absolute, uncompromising commandment that gives them clear prescriptive guidelines. Instead, physicians must *share* personal responsibility for the ethical bases and decisions underlying their actions. The new ethic forces physicians to engage their patients and their families in dialogues concerning issues dealing with the practical evaluation of life, suffering, death, departure from social ties, economic considerations, and unfulfilled promise. Put differently, the emerging system of medical ethics requires physicians to share the management of uncertainty with others who are not members of the professional medical epistemic community.

From our perspective, this has profound consequences for both the authority and the autonomy of that epistemic community. Specifically, it means that the medical community is forced to acknowledge that its task is less one of uncertainty reduction than of the management of uncertainty, and that the absorption of risk is increasingly accompanied by the sharing of risk. This shift away from claims that imply risk reduction to more realistic acknowledgement that medical professionals can do no more than assist in the management of uncertainty and risk tends to undermine the previously absolute authority of the profession in dealing with matters of life and death. Even more importantly, the obligation to consult others—such as the patient and family members—and include their ethical systems in the decision, as well as the obligation to admit the medical significance of ethical considerations structured in other sectors of the society, further undermines the autonomy of the medical community. In short, the technological development of new controls over life and death has made an increasingly articulate and educated public, as well as the medical profession itself, outspokenly reluctant to permit physicians to make God-like decisions about the continuation or termination of existence alone.

Those considerations would seem to augur the end of a certain period of social history in which specialized, technical knowledge and expertise are granted absolute authority when applied by professional epistemic communities that have complete autonomy over the subject matter with which they work. In other words, these developments suggest that bodies of specialized, technical knowledge can no longer be granted complete and unrestrained autonomy from emerging models of common sense and, therefore, that the authority of practitioners belonging to professional epistemic communities can no longer be exercised in complete independence of the preferences and values of their clients. Indeed, we suggest that medical diagnosis and treatment will increasingly come to be the result of negotiated agreements, not only between the client-receiver and professional giver of services, but also involving friends, relatives, and interested third parties, including representatives of formal organizational interests. In addition, it seems safe to suggest that this emerging pattern involving the sharing of crucial decision making in professional arenas both results from and serves to increase the sharing of technical expertise previously monopolized exclusively by the professional community.

As these developments undermine the special moral authority, sociocultural autonomy, and somewhat charismatic status that have surrounded professional practice, an emerging trend toward consumerism generates organized group pressures to have health provider decisions checked on a massive scale. The new demands for accountability associated with this emerging consumerism represent another direct challenge to the medical autonomy, which has heretofore defended physicians' freedom from oversight. This has led to the development of Professional Standards Review Organizations that in effect monitor doctors' standards of performance. The manner in which all of these developments serve to limit the autonomy and authority of professional epistemic communities is clearly illustrated by the growing demand that professionals should be held accountable to the public for their actions, not merely to their peers. The fact that this call for public accountability has been taken up by broad segments of the society, as well as by the political system itself, implies that the legitimacy of the authority and autonomy of the medical epistemic community is no longer widely acknowledged and can no longer be exercised in complete disregard to emerging patterns of morality and, especially, post-modern common sense.

CONCLUSION

We will return to the issues raised here in the context of the following chapters. These illustrations conclude our sketch of the idea of social structure in relation to knowledge; they also show the complex dynamics in the intertwining between cognitive activities, epistemic criteria, and social structure. We have thus introduced our major conceptual tools and analytical imagery in terms of which we will proceed. The following part of this book deals with the knowledge system and its components more explicitly. We begin with an outline of the organizing scheme.

PART

II

Bibliographic Notes

The two chapters that comprise Part II of the book attempt to develop a conceptual framework and sociological perspective for the analysis of post-modern knowledge. In addition to a number of the central works referred to in the bibliographic notes to Part I of this book, our perspective draws heavily on two related formulations: Susan Langer's *Philosophy in a New Key* (1951) and Ernst Cassirer's *An Essay on Man* (1944). The discussion of "knowledge as a sociological construct," as well as the arguments concerning social structure as constituting a knowledge-containing system or framework for knowledge, draw heavily on George Gurvitch's work on *The Social Frameworks of Knowledge* (1972). We are also indebted to the works of Claude Levi-Strauss, especially his 1963 statement in *Structural Anthropology* in the translation by Jacobson and Schoepf. Two important recent contributions which bear indirectly on the subjects considered in Part II are Peter M. Blau's (1974) Presidential Address to the American Sociological Association entitled "Parameters of Social Structure" and Robert K. Merton's collection of essays entitled *Sociological Ambivalence* (1976). Our attempt to illustrate our underlying perspective through an examination of "sleep and the paramount reality of everyday life" draws heavily from an imaginative article by Vilhelm Aubert and Harrison White entitled "Sleep: A Sociological Interpretation" (1959).

Chapter 5 discusses a number of concrete topics involving contemporary professions in order to illustrate some specialized work communities that develop structural frameworks for technical knowledge and expertise. A number of medical trends and specialties are considered in some detail. The most important influence on our interpretation of a wide range of events in the medical domain—as well as one of the central influences on our conceptual framework in general—is Eliot Freidson's *Profession of Medicine: A Study of the Sociology of Applied Knowledge* (1970). Indeed, the influence of Freidson's writings can be found in almost every chapter of this book. Although somewhat more limited to its applicability, the work of labelling theorists has exerted significant impact on our treatment of issues in the medical and mental health domains, especially Thomas J. Scheff's *Being Mentally Ill: A Sociological Theory* (1966).

In our treatment of the interplay of professional ideology and professional identity as cornerstones of the specialized psychoanalytic community, the two volumes by William E. Henry, John H. Sims and S. Lee Spray and entitled, respectively, *The Fifth Profession* (1971) and *Public and Private Lives of Psychotherapists* (1973) have been especially useful sources of empirical examples supported by data. Our discussion of recent changes in American medicine as an illustration of the technological transformation of various professional communities derives from a number of studies. Among the more important works that document these changes and provide support for our interpretations are the following: R. M. Magraw's *Ferment in Medicine: A Study of the Essence of Medical Practice and of its New Dilemmas* (1966); J. B. Richmond's *Currents in American Medicine: A Developmental View of Medical Care and Education* (1969); the pioneering article by Donald C. Ransom and Herbert E. Vandervoort entitled "The Development of Family Medicine: Problematic Trends" (1973); and the article by Charles A. Janeway entitled "Family Medicine—Fad or for Real?" (1974). Finally, the most powerful analysis of the changing moral bases of modern medical ethics, as well as of the emergence of new professional definitions of the increasingly salient and problematic medical situation of dealing with the terminal patient can be found in the article by Talcott Parsons, Renee C. Fox, and Victor M. Lidz entitled "The Gift of Life and Its Reciprocation" (1972).

PART
III

The Knowledge System —
An Analytical Sketch

In the third part of this book, we turn directly to the main aspects of the knowledge system. We begin with chapter 6 on knowledge production, which deals with the creation or discovery of new knowledge. Our analysis emphasizes the role and historical development of scientific, scholarly, and professional epistemic communities as strategic sites for the institutionalization and then bureaucratization of scientific and professional epistemic communities and raises questions about the likely directions of future development.

Chapter 7 deals with the crucial linkages and intermediaries in the post-modern knowledge system; namely, the process of knowledge organization, dissemination-distribution, and storage-retrieval. These connective processes and functions are essential, since without them it would be meaningless to talk about the existence of a knowledge system. Moreover, although other writers have discussed processes of knowledge production and utilization, albeit from different perspectives than employed here, few of these analyses have devoted systematic attention to the structural linkages and channels that have emerged in contemporary society to connect knowledge production to utilization and application. This chapter naturally strains against its boundary. The web of these mediating linkages and channels is of such scope and magnitude that we can only provide hints and points of departure for further analyses, rather than do full justice to the complexity of the subject.

Chapter 8 concludes Part III with an analysis of post-modern knowledge application. This chapter explores relations between the knowledge system and the manifold worlds of practical activity. We offer a conceptualization of the act of knowledge application that emphasizes its extraordinary characteristics: it represents a break with routine, habitual behavior. The implications of this view of knowledge application lead to some rather unusual interpretations of familiar events. With this chapter, our analytical scheme for examining the knowledge system comes full circle in discussing the incorporation of specialized, technical knowledge into the routines of everyday social life.

An Outline of the Knowledge System

This brief section outlines one of the main organizing schemes we will use in the remainder of the book: the idea of the social system of knowledge. One can analyze the complex reality of social life with special attention to a variety of aspects. For example, one may be interested in power, or in the channels that social structures provide for the flow of communications, and the like. Analytical constructs such as the ideas of the power structure, or the communications network, or the economic system and others guide such modes of dissecting social reality analytically. Each of these modes of abstracting from reality is in some fashion interdependent with the others, yet each also is analytically distinct and provides a conceptual focus of attention. In analogy to these conceptions, we use the term knowledge system to refer to the particular view that emerges when one studies the social distribution of knowledge-related processes.

We use the conception of a societal knowledge production and use system as an organizing device. For example, one could classify, describe, and explain the differences in the degree of differentiation in the arrangements for social knowledge functions. In societies in which knowledge is not codified, but transmitted through the learning of skills embedded in social roles, taking the form of wisdom, the knowledge system does not appear as an empirically separate aspect of social structure. In post-modern society, however, the knowledge system is most definitely emerging as an empirically differentiated, even deliberately planned, set of institutions and roles.

We can distinguish between different, but interrelated, social activities in the knowledge system. First, there is the creation of new knowledge that we term knowledge production. *Second, there is the* organization of knowledge *into coherent bodies. Third, there are the activities and institutions of* knowledge storage; *next are those of* knowledge distribution and accessing. *Finally, we can distinguish between* knowledge application, *putting knowledge to use in solving of practical problems, and* implementation, *absorption into everyday routines of explicitly knowledge-based plans for action. Clearly, these distinctions refer to interdependent and often fused aspects of the empirical knowledge system. It would be erroneous to argue that knowledge production is independent of the other functions; indeed, as we will see later, knowledge application particularly may be intertwined with production processes. Yet, we consider it helpful to sketch the structural location and dynamics of knowledge processes or functions in the social system.*

How the functions of knowledge production, storage, distribution, and use or application can be carried out depends on the existence of cultural models explicated to facilitate accom-

plishing the relevant task. This means that the knowledge system of a society, once structurally differentiated, is particularly dependent on institutional innovations and inventions. The knowledge system of post-modern society changes rapidly and appropriate new roles must be created and separated just as rapidly from the older clusters in which they were embedded. For example, with the emergence of highly formalized mass systems for knowledge distribution, journalists become specialized as science reporters and function as middle men between research and knowledge use, experts in knowledge translation and legitimation.

The major innovative development in the knowledge system was the development of institutional patterns for academies and universities. Universities have not always conceived of themselves primarily as institutions for knowledge production. In many periods of their history they gave much more emphasis to the functions of knowledge storage and transmission. However, they were always central institutions in knowledge systems. They were and continue to be structurally tied to the second major cluster of cultural models for knowledge functions; namely the professions and the scholarly disciplines. In these we find culturally defined and normatively sanctioned organizational arrangements that focus either on knowledge use for practice, as in the professions, or on knowledge production and growth, as in the academic disciplines.

The complex institution of the modern university is one of the central features of the post-modern knowledge system. University linkages with the professions provide many of the connections that maintain a minimum level of integration in the system. However, we will see that there are many other cultural models for knowledge processes, some of which emerged during debates as to how knowledge is best put to use or how scientific research can be responsibly governed.

These remarks should suffice to indicate our organizing schema and the way we approach the idea of the knowledge system. Yet one further point needs to be made. Much of the recent literature that discusses knowledge production and use is instrumental and managerial both in style and tone. This is most understandable in that these works address problems of considerable urgency in the contemporary world, in which the improvement of social conditions through organized knowledge use are important matters that challenge the ingenuity of knowledge workers, demanding that they provide practical recommendations and remedies. Much of the literature that deals with planning for innovation, purposeful change, and intervention in practical affairs merely attempts to translate knowledge-based inventions into social techniques. We respect the purposes of these authors, but ours are different. We write from what one might call a naturalistic stance; sorting, describing, and exploring. From this point of view, we find much of merit and even more of interest in this literature and occasionally refer to it, particularly in the chapter on knowledge applications. However, we are concerned with connections that this literature overlooks or ignores, for example, the link between social movements and the knowledge system as a whole. Our theoretical orientation is that of the observing sociologist, not that of the manager or planner.

6

Knowledge Production

In a sense it is arbitrary to begin our treatment of the components of the knowledge system with an examination of knowledge production. We might as well have begun with the study of knowledge use. However, there is some plausibility in examining first the growing edge of the system, the creation of historically new knowledge. We will see that many of the considerations important in this domain are significant in the other domains of the system as well.

THE GROWTH OF KNOWLEDGE

Contemporary common sense takes many things for granted that on reflection appear quite extraordinary. It is necessary to remind ourselves of the historical recentness of the social and cultural changes that are the bases for the seemingly inexorable growth of knowledge in the contemporary world. The formal professionalization and institutionalization of science did not begin much earlier than the nineteenth century. The prototype of the research university emerged in nineteenth century Germany, followed shortly by the enormously consequential innovation of the university-related research institute. It was only after the Civil War that graduate and professional schools in the American university began to adopt the departmental structure that subsequently provided the institutional framework for academic discipline-based research. The idea that the production of new knowledge could not be left as the dignified pursuit of wealthy, educated, and reflective gentlemen of leisure, nor could be considered an activity dependent on the fortuitous emergence of insightful, creative genius, but that it must become collectively organized and routinized work, pursued in specifically designed institutions, was an historic innovation of profound significance.

It certainly is true that the development of science predates this period and its innovations. Obviously scientific modes of inquiry existed in earlier periods of Western history as well as in other civilizational complexes. However, the realization that scientific knowledge grows cumulatively, and that its growth is affected by the manner in which scientific activity is organized, arose uniquely in the Western world. These new conceptions of knowledge production were preconditions for the more specific realization that work organizations, educational programs, and professions could be designed for systematically developing and applying knowledge. In fact, the recognition of these possibilities supported a steadily increasing expansion of organized knowledge production during the late nineteenth and early twentieth centuries. Whether one considers the frequency of scientific discoveries, or counts the number of employed scientists, or reviews the expansion of specialized educational structures designed to produce and disseminate knowledge, the picture of rapidly expanding organizational frameworks constructed specifically for the production of specialized, technical bodies of knowledge remains the same.

Price (1961) has observed that scientific productivity increases exponentially. For example, he examined the number of scientific journals from the seventeenth century to the present and concluded that this population increased by a factor of ten every fifty years beginning in the middle of the eighteenth century. It appeared to Price that every scientific advance became the basis for several further advances so that science diversified as it grew. It is difficult to imagine that such a rate of growth can continue indefinitely; yet new forms of structuring knowledge and the complex differentiation of what is known do not provide us with the bases for predicting inherent limits to knowledge growth in science.

However, the explosive growth of knowledge is not characteristic of its scientific core alone. Codified professional knowledge and
systematic expertise have grown in a similar fashion. These bear complex relationships to the development of scientific understandings,
but it does not make sense to call all professional knowledge, which
is taken as factual, scientific. For example, there has been considerable development in the role of expert knowledge concerning child
rearing. Much expert advice can be shown to be devoid of a scientific
knowledge base; nevertheless, it may be *taken* as professionally reliable knowledge. Blumenfeld (1976) describes the rise of child-rearing
expertise and the successive developments that follow from increasingly explicit and codified knowledge bases in this area.

Both scientific bodies of knowledge and codified expertise
related to science have expanded at an accelerating rate since the two
world wars. For example, the creation of large scale institutes for
social research was essentially a post-World War II phenomenon. So is
the rise of so-called big science, requiring extremely costly scientific
apparatus of great complexity and gigantic size. One need only think
of accelerators, or radio astronomy installations to make the point.

Indeed, knowledge production has become one of the key
issues in public policy. Concern with the expenses and resources
necessary for research is as important as concern with the consequences of new knowledge and technology. During the period of
industrialization and modernization, it became clear that policy formation had to consider specifically the certainty of change. In the
post-modern era, even more complex issues of the relation between
knowledge and change are posed. The effort to forecast technology
for policy purposes underscores this point.

In the following discussion of knowledge production, we will
focus on specialized social contexts for the development of new
knowledge. We will emphasize the relationship of these contexts to
larger social structures in which they are embedded and illustrate the
dynamics of rationality and epistemic criteria in them. In this connection, we argue that rationality itself is not extrasocial in nature.
Its pure crystallizations in mathematics, logic, and theory are themselves achievements of social systems. Therefore, our purpose in this
chapter is to show both the embeddedness and the specialized
autonomy of rational, systematic knowledge production in order to

lay the groundwork for subsequent considerations of knowledge distribution and application. We will proceed from a discussion of certain social frameworks for specialized knowledge to a consideration of scientific differentiation and the dynamics of knowledge growth internal to specialized epistemic communities. Our emphasis on the embeddedness of specialized knowledge production lays the groundwork for themes to be raised later.

SOCIAL FRAMEWORKS FOR SPECIALIZED KNOWLEDGE: GURVITCH

In the broadest sense, social structures are always knowledge containing systems in that they are simultaneously determined by social categories and positions; orientations, knowledge, and belief systems; and the reference frames of their occupants. Taking this rather obvious premise seriously leads one to appreciate how social structures become frameworks for inquiry. Indeed, we will argue that the social contexts and embeddedness of even highly specialized inquiry have profound significance for the form and content of the knowledge generated in that they define the characteristics of the sociocultural anchorages and resources available to it.

The nature of social frameworks for knowledge has been explored in considerable detail by George Gurvitch (1972) who described the "functional correlations" between various types of knowledge and the social frameworks within which they exist. The social frameworks Gurvitch (1972:16f) treated range from macrosocial structures like global societies, social classes, and status groups to microsocial elements of the immediate surroundings of a particular investigation. Gurvitch then related these frameworks to several different types of knowledge.

Gurvitch started with basic perceptions of the external world; following Durkheim, these are treated as social facts. This emphasizes that perceptual knowledge of the external world is dependent on collective constructions of time and space that provide an encompassing framework for knowledge and/or inquiry. Social knowledge, which he describes as "knowledge of the other, the we, groups, classes, and society," is rather special because it is in part constitutive of social structure themselves. (This is a point with which we are in emphatic agreement.) Gurvitch then proceeds to discuss common sense as a special type of social knowledge (Ibid.:27f):

> It consists of a combination of knowledge of the other and the we; of perceptual knowledge of the external world . . . ; certain simple forms of technical knowledge—physical techniques; polite manners, and maintaining reserve and distance. It tends to favor traditional knowledge, that of older and experienced people, and *savoir vivre*, which neither children nor more generally the younger generation possess.

Thus, Gurvitch suggests that common sense knowledge possesses an integrative function. However, he seems to have overlooked a number of major transformations in common sense brought about by the emergence of post-modern society with its pervasive insistence on knowing how to gain access to diverse bodies of knowledge and sources of expertise.

Gurvitch argues that the central feature of technical knowledge is that it is directed toward effective manipulation. Even though such knowledge is subordinate to other goal structures, it frequently tends to become independent of these goals, and is valued in and for itself (Ibid.:28f). In contrast, political knowledge is structured around problems of policy; this type of knowledge includes both an element of strategy and an element of reflection as components for the guidance of political action.

Gurvitch treats science and philosophy as types of knowledge that are unusually detached from concrete social frameworks. They are social creations, but their social groundings and contents transcend specific historical forms. Gurvitch characterizes philosophy as "a second order knowledge" that is "grafted retrospectively on other kinds of knowledge" and strives to integrate these partial aspects of knowledge into infinite totalities, so as to validate them. (In this latter characterization, Gurvitch's conception of philosophy bears some resemblance to Geertz' interpretation of "Religion as a Cultural System" (1966). In philosophical knowledge, specific sociohistorical collective reference frames may be transcended, Gurvitch argues, although even there they have some impact on the body of knowledge.

Gurvitch's discussion of scientific and philosophical knowledge serves as an important reminder that, while specialized knowledge of all kinds can and frequently does partially transcend the specific circumstances of its creation, the sociology of knowledge must consider the contextual embeddedness—as well as the transcendence—of knowledge and, especially, of the organizational frameworks in which it is produced. From the perspective of social systems, specialized knowledge production emerges out of broad contexts and can be seen as a component of their collective learning capacities. However, knowledge appears to be under encompassing constraints of rationality as a collectively emerging phenomenon. Further pursuit of these themes would take us beyond the confines of this chapter. Let us consider in more detail how certain forms of specialized knowledge emerge from previously undifferentiated contexts.

THE EMBEDDEDNESS OF SPECIALIZED INQUIRY: FOUCAULT

It is instructive to follow the manner in which a specialized form of knowing emerges and becomes differentiated from an as yet undifferentiated contextual matrix. Foucault refers to any specialized domain of knowing as a "discursive formation." Some of his illustrations refer to the emergence of the conception of "madness" in France. The idea of madness is not simply a phenomenon the meaning of which is given in itself. The very possibility of forming a psychiatric perspective and constructing the analytic objects of psychiatric investigation required the prior establishment of a domain of discourse within which such analytical objects could emerge. The understanding of madness is determined by diverse institutional arrangements, social forces, and cultural patterns. Thus, the working relationships between the insane, their guards, and the physicians who treated them had a great deal to do with the way in which psychotherapeutic conceptions emerged. These working relationships were, in turn, embedded in family relationships with varying ability to tolerate and/or encapsulate deviance. These family relationships were also embedded in more secondary associations, such as occupational and religious communities that, in turn, were embedded in still larger sets of relationships. All of these social ties and networks influenced in complex fashion the emerging classification of a person as mad or sane. Foucault suggests that there are "authorities of delimitation" that, in the case of emerging conceptions of madness, were those groups and agencies in society empowered to authoritatively delimit and designate madness as an object of inquiry. In the late eighteenth and early ninteenth centuries, medicine emerged as the dominant authority of delimitation, although it was certainly not the only one. The law and legal system, and intellectual conceptions of authority also had important influences in delimiting the emergent concept of madness as an object of discourse and a domain of inquiry.

These arrangements work in a highly specific way through networks of referral, through performance of social measurements, and through classification of behavior patterns of persons, so that an institutional complex emerges—including asylums as well as organized professions—that works on this reality called madness. These institutional contexts gradually come to be differentiated clearly and distinctly from the surrounding society as they acquire functional specificity. In Foucault's terminology, "grids of specification" have begun to emerge. (We have been using the terms *frames of reference* and *epistemic criteria* to describe this increasingly specialized cognitive template that develops for dealing with an object or domain of inquiry.)

The dynamic contexts in which social structures and their cognitive concomitants emerge contain embryonic forms of novel and specialized modes of inquiry. Foucault emphatically stresses the complex systemic interdependence among the various factors: a specialized body of knowledge, a domain of discourse in the process of formation, cannot simply be described as a set of objects that are successively discovered and classified, so that theories can then be formed about them. On the contrary, there is a reciprocal interpenetration and interdependence between an emergent mode of discourse within a specialized domain of knowledge and the objects to which it relates. "It would be quite wrong to see discourse as a place where previously established objects are laid one after another like words on a page" (Foucault, 1972:42f). In the case of madness, one can see how the emergent mode of discourse shaped the behavior of families, legal procedures, courts of law, and in turn the mentally ill themselves. As Foucault puts it (Ibid.:44):

Let us generalize: in the 19th century, psychiatric discourse is characterized not by privileged objects, but by the way in which it forms objects that are in fact highly dispersed. This formation is made possible by a group of relations established between authorities of emergence, delimitation, and specification. One might say, then, that a discursive formation is defined (as far as its objects are concerned, at least) if one can establish such a group; if one can show how any particular object of discourse finds in it its place and law of emergence; one can show that it may give birth simultaneously or successively to mutually exclusive objects, without having to modify itself.

It is possible to expand Foucault's conception and relate it to the definition of social problems and problem domains for inquiry in contemporary society. A very brief example serves to illustrate Foucault's formulation.

In analogy to Foucault's analysis of the emergence, definition, and classification of madness in nineteenth-century France, consider the definition of crime in contemporary society. There is no doubt that crime and criminality exist as hard and constraining facts, not easily subject to manipulation despite many attempts in that direction. However, exactly what crime is, and how it is known and talked about, can be conceptualized in analogy to Foucault's approach. Moreover, the establishment of the uniform crime statistics by the FBI represents the construction of quantitative social indicators that delimit and define a domain of discourse for both policy formation and rational inquiry. Institutional arrangements, the interests of various participating organizations, the classificatory schemes available, and the mode of discourse itself, interact to produce a complex reality that then becomes the subject of a body of knowledge. At the same time, such an emergent body of knowledge is reactive and interrelates with, as well as modifies, the social processes it purports to describe and explain.

These interrelated factors that produce, sustain, and are influenced by an emergent mode of discourse together comprise what Foucault calls a "positivity." The term is not used simply to describe a set of statements or a body of knowledge as such, but to describe the specific social contextual forms of knowledge accumulation within an as yet unreflected institutional domain. Social problems may be constructed primarily on a moral basis, through the repeated observation of demonstrable discrepancies between actual and morally desirable states of affairs. The articulation of a social problem tends to occur much in the manner in which Foucault has described the emergence of a positivity. These may be conceived of as emerging islands of knowledge and inquiry that are only gradually integrated with other domains through formal-logical, substantive or methodological convergences.

There are naturally different trajectories for the historical emergence of a domain of knowledge and inquiry—what Foucault calls a "discursive formation"—and it is impossible to describe a unitary and simple pattern or trajectory across several fields. But Foucault adds the idea of different thresholds and their chronological sequence as a means to avoid examining an infinite number of empirically divergent trajectories. He speaks of the "threshold of positivity" when a domain of inquiry achieves "individuality and autonomy"; that is, when it is recognized by both practitioners and inquirers as a bounded area of and for specialized, technical inquiry. The core of such a domain is usually quickly appropriated by some specialized or specializing work community. Thus, a domain that has achieved individuality and autonomy, as well as recognition as a bounded area of inquiry, will be an arena of contests for appropriation. In other words, professionalization is likely to set in when a specialty is recognized as such and a chartered domain of discourse has been authoritatively allocated to some collectivity. Foucault describes the developments in this sequence as follows (Ibid.:186f):

> When in the operation of a discursive formation, a group of statements is articulated, claims to validate (even unsuccessfully) norms of verification and coherence, and when it exercises a dominant function (as a model, a critique, or a verification) over knowledge, we will say that the discursive formation crosses the threshold of epistemologization.*

In this statement, Foucault indicates that further development of such a domain, his "discursive formation," must proceed reflectively. That is, there is a tendency toward the codification of procedures and the explicit formulation of epistemic criteria.

An explicit concern with reflectivity per se introduces important alterations into a system. Before awareness or reflectivity emerges, observed, experienced, or recorded facts are, almost of necessity, simply accepted as such; they are not selectively evaluated. "Epistemologization" introduces the critical reflection that is necessary for sifting fact from error in relation to the community's guiding

*Michael Foucoult, *The Archeology of Knowledge* (tr. A.M. Sheridan Smith). © 1972, Pantheon Books, a Division of Random House, Inc.

epistemic criteria. This is likely to be a turbulent process fraught with conflict, since it typically leads to the modification or even rejection of the working knowledge of practitioners or inquirers prior to epistemologization. Implicitly accepted reality tests are replaced by explicit and more formal epistemic criteria, with the result that systematic, specialized, frequently scientific or critical knowledge comes into direct conflict with both common sense and pragmatically validated working knowledge. This can be reformulated in terms of the concepts introduced in the third chapter during the discussion of reality tests. Specifically, prior to epistemologization a discursive formation is likely to employ implicitly accepted pragmatic reality tests or reality tests that rely on consensual validation, in order to assess the validity and reliability of statements or observations that are presented as facts. Epistemologization means that the specialized work community begins to codify, formalize, and articulate the reality tests on which it relies most heavily and begins to substitute scientific or formal-logical and critical reality tests for former tests based on pragmatic workability and common sense consensual validation. Numerous examples of this process can be found throughout medicine in which critically and systematically gathered information invalidated dominant beliefs and prevailing practices based on earlier, pragmatic reality tests. The practice of bloodletting and the fate of homeopathy illustrate this process.

When the epistemological criteria and the methodological rules of inquiry comply with canons of empirical investigation and rules for forming theoretical propositions, Foucault speaks of crossing the threshold of "scientificity."

> And when this scientific discourse is able, in turn, to define the axioms necessary to it, the elements that it uses, the propositional structures that are legitimate to it, and the transformations that it accepts, when it is thus able, taking itself as a starting point, to deploy the formal edifice that it constitutes, we will say that it has crossed the threshold of formalization.
>
> *(Ibid.:187)*

Thus, Foucault sees formalization as emerging in the course of this process; it means the crossing of another threshold that again brings about a qualitative change in the mode of discourse. As we have already cautioned, however, it would be a complete misunderstanding to interpret Foucault as suggesting that these thresholds necessarily must be passed in a particular invariant sequence. Foucault makes it clear that it is not possible to construct a rigid sequence or order in which each of the thresholds is crossed. On the contrary, Foucault proposes these thresholds as analytical reference points for studying the emergence of inquiry or, in our terminology, the emergence of the organized production of systematic knowledge.

THE EMERGENCE OF AUTONOMY IN SCIENCE

The decisive break from classical and scholastic learning was also a break that meant the beginning of autonomy in inquiry. This growth of inquiry into what has become the *modern temper* began between the mid-sixteenth century—in 1543 both Copernicus' and Vesalius' major works were published—and the early seventeenth century, the century of genius, which saw the work of Kepler, Galileo, Bacon, Descartes, and Harvey among others. Substantively, this break rested on the assumption that the truth is capable of being discovered at a particular moment and that it thus is capable of being sharply revised. The *new learning* rested on a faith in critical thought, inquiry, and experience as opposed to traditional authority and revelation. The assumption that knowledge could be reappraised and new discoveries made that would constitute increasingly valid successive approximations to the truth involved the substitution of empirical reality tests for religiously based authoritative ones.

The insurgent new learning quickly found itself in conflict with the classical, scholastic traditions associated with the Church. The trial of Galileo in 1632 initiated the separation of experimental knowledge and the Roman Church, but it did not halt the spread of critical inquiry, even in Italy. Rather, it sharply defined the issues so that the new knowledge formed itself as a countersystem to the old. It rapidly shifted from Latin, the language of established learning, to the vernacular. Its naturalistic terms of explanation contrasted with supernaturalism. Its current, disciplined observation contrasted with traditional, authoritative interpretations of the ancient texts, resting on either revelation or intricate deductive schemes.

From our perspective, perhaps one of the most interesting facts about the emergence of this early modern system of knowledge is that it took place in socially interstitial areas, largely outside the universities that were dominated by religious authorities and the scholastic tradition. The new scientific and philosophical investigations did not significantly enter European universities until the early nineteenth century. Moreover, they did not enter first the older, respected, self-governing universities in England or France, but the new, state-supported Ecole Polytechnique in France and the newer, governmentally established universities in Germany. From the mid-sixteenth century until the nineteenth century, then, science and the new learning had the independence and autonomy, but also the marginal existence, of self-support. The early innovators were amateurs whose studies, although often full-time, were largely unpaid and whose work was undertaken outside of traditional formal institutions.

One mechanism of central importance in early as well as in contemporary science, as revealed in the importance of proper communication and the stress on the reliability of research results, is the reciprocal exchange of information among scientific workers. The early scientific societies that institutionalized this exchange were simultaneously secular epistemic communities and cultural free spaces. Just as the main support of research through the seventeenth century came from investigators themselves, so the early scientific societies—such as the Royal Society of London—were entirely self-supported by membership fees, gifts, and bequests (Krohn, 1972:34–43). The rise of amateur science rested fundamentally on the establishment of these self-supporting, specialized scientific societies. In England, for example, no less than twelve such societies began between 1788 and 1839 without the serious establishment of science in any of the major English universities.

Our argument, then, is that the emergence of modern knowledge did not require widespread secularization or an end to the religious authority of knowledge as truth, but rather the emergence of socially defined cultural pockets or free spaces within which epistemic communities of amateur researchers were able to establish sufficient autonomy to proceed with their investigations. This was possible precisely because the new knowledge was viewed as marginal to society. "When not seen as heretics or critics of traditional learning, scientists were seen as cranks, mere hobbyists, or wanderers" (Ornstein, 1928:122). As Krohn observes (1972:66): "Science was first able to be an open, progressive system of knowledge precisely because it was so socially marginal. . . ."

In these ways, the early scientific societies appear to have combined the critical attributes of both epistemic communities and cultural free spaces. That is, these societies certainly comprised specialized workers committed to a specific observational and experimental criterion concerning valid methods of obtaining knowledge. At the same time, however, these societies also encouraged experimentation with new methods and techniques in order to obtain reliable knowledge, even if that knowledge contradicted truth taken as established. The fact that the early scientists were largely self-supporting amateurs also appears to have contributed to the acceptance of their societies as cultural free spaces within which statements or findings could be presented to a nondefensive audience. It was when their views or discoveries were disseminated beyond these boundaries of cultural free spaces that the Church moved rapidly and repressively to punish workers who failed to perceive the fragile and conditional nature of their scientific free spaces. The trial of Galileo in 1632 forcefully demonstrates that.

SCIENTIFIC NORMS AND DYNAMICS OF KNOWLEDGE GROWTH

Within the autonomous domain of emerging and gradually institutionalized scientific knowledge production, a normative structure developed that, in large measure, regulated scientific activity and became a major determinant of the dynamics of subsequent knowledge growth. One of the most significant contributions to understanding the normative guidelines that circumscribe scientific research grew out of Robert K. Merton's (1957) studies of the relationship of science to Protestant religious orientations and economic changes in seventeenth-century England. In the course of examining selected periods in the history of science and scrutinizing its problematic growth in sociopolitical systems selectively hostile to it—most notably, the partial repression combined with selective support of science in modern totalitarian regimes—Merton attempted to articulate the central ethos of modern science (1957:552f):

> The institutional goal of science is the extension of certified knowledge. The technical methods employed towards this end provide the relevant definition of knowledge; empirically confirmed and logically consistent predictions. The institutional imperatives (mores) derive from the goal and the methods. The entire structure of technical and moral norms implements the final objective. The technical norm of empirical evidence, adequate, valid and reliable, is a prerequisite for sustained true prediction; the technical norm of logical consistency, a prerequisite for systematic and valid prediction. The mores of science possess a methodological rationale but they are binding, not only because they are procedurally efficient, but because they are believed right and good. They are moral as well as technical prescriptions.

Merton then proceeded to delineate four central institutional imperatives of science: universalism, communism, disinterestedness, and organized skepticism. By universalism, Merton referred to canons of truth that are objective and impersonal, and are exempt from the particularistic claims of ethnocentrisms and nationalisms. He used the term communism to describe the normative requirement that all substantive scientific findings belong to the entire professional scientific community of workers and are public in nature. The institutionalized norm of disinterestedness requires that the diverse motives of individual scientists are subordinated to the moral as well as institutional norms of the scientific enterprise. For example, Merton observed that the virtual absence of fraud in science can hardly be attributed to unusual moral integrity on the part of scientists as individuals. Instead, he argued that the "translation of the norm of disinterestedness into practice is effectively supported by the ultimate accountability of scientists to their compeers. The dictates of socialized sentiment and of expediency largely coincide, a situation conducive to institutional stability" (Ibid.:559). Finally, the institutionalized norm of organized skepticism requires continuous scrutiny of scientific assertions in terms of both logical and empirical criteria.

Merton's conception of the normative ethos of science expressed only one side of the matter: that is, the norms he described are unquestionably held by working scientists, but not in so uncomplicated a fashion. Merton himself demonstrated that the reward structure of science—for example, the enormous significance given to priority in discovery or innovation—provides incentives for originality and further innovation. He subsequently came to the realization that the structure of a scientific role may consist of "a dynamic alternation of norms and counternorms" (Merton and Barber, 1963: 91–120). On the basis of this insight, Merton has come to conceive of the dynamics of science as involving a dialectic among sets of norms in tension with each other. Mitroff (1972 and 1974) produced empirical support for this conception through a detailed study of prestigious scientists involved in the study of lunar rocks acquired during the Apollo moon landings. Specifically, although the normative commitments and constraints of science emphasize rationality, disinterestedness, universalism, and their concomitants, Mitroff (1972) was able to demonstrate counternorms of intense commitment, single-mindedness of purpose, particularism, limited secrecy, and organized dogmatism in persevering adherence to a specifically favored theory. It is in the interplay between the institutionalized norms that regulate science, the intense personal commitments to particular theories and research programs, and the structure of research careers in the academic-scientific disciplines, that one finds the sociocultural dynamic of institutionalized science.

The discussion so far has emphasized the significance and consequences of normative and structural constraints and influences on disciplines, using scientific research as the paramount model for illustration. However, it is also necessary to consider the dynamics of knowledge production and growth that derive from the internal symbolic structure of knowledge itself.

Thomas Kuhn's (1962 and 1970) analysis of "the structure of scientific revolutions" argues that the organization of scientific knowledge itself gives rise to the perception of certain puzzles. More specifically, at any given point in time a science may be dominated by a specific paradigm or fundamental image of a science's subject matter. The paradigm is based on an exemplar of particularly compelling nature. It serves, among other things, as a mode for organizing the coherence of a scientific body of knowledge. Normal science is a

period of knowledge accumulation in which scientific work progresses in terms of the reigning paradigm—filling in gaps in the paradigm or expanding it. Inevitably, such work spawns anomalies or
puzzles that cannot be explained within the existing paradigm. If
anomalies continue to mount, a crisis stage is reached, which may
result in a scientific revolution in which the reigning paradigm is
overthrown and some challenger takes its place at the center of the
science. At this point, a new cycle is begun that will end with the
defeat of the new paradigm. Kuhn argues that the period of revolution contains significant noncumulative changes in the status, subject
matter, and definition of science. The key concept in Kuhn's model
and in analyses based on it is the elusive notion of paradigm. Kuhn
(1970:175) originally proposed a very general definition of paradigm
that embraced "the entire constellation of beliefs, values, techniques,
and so on shared by the members of a given community." In a later
edition, Kuhn (1970:175) proposed a much narrower definition of
paradigm as "the concrete puzzle solutions which when employed as
models or examples can replace explicit rules as a basis for the solution of the remaining puzzles of normal science."

Several significant points relevant to the systematic production of new (scientific) knowledge are embedded in Kuhn's model.
From Kuhn's perspective, knowledge growth does not occur in a
straightforward, incremental, linear fashion; scientific development
cannot be thought of as simply additive. We have already encountered support for this view in Foucault's ideas concerning discontinuities and thresholds. Kuhn's work also emphasizes the significance of
very basic images or models for ordering knowledge—master models—
that provide a sense of consistency in knowledge production efforts,
organize available information, and aid in the location of problems.
The nature of problem identification in times of normal and revolutionary science is quite different. Normal scientific inquiry involving
the careful and painstaking addition of small increments of knowledge differs drastically from the extremely risky, but also extremely
rewarding—if successful—attempt to reorganize an entire body of
knowledge in a scientific revolution. Finally, Kuhn (1962) views the
emergence of a new paradigm as analogous to political phenomena. A
paradigm wins out over another because its advocates succeed in
being more persuasive. It may not necessarily be the case that the
new paradigm is better than its rivals. In an article written in

response to his critics, Kuhn (1974:261) argued that one paradigm replaces another for good reasons, including "accuracy, scope, simplicity, fruitfulness and the like." This means that Kuhn is turning to a more rationality-oriented conception of scientific revolutions in which good reasons, inhering in the internal symbolic structure and applicability of a paradigm, have replaced both nonrational, irrational, and political factors in explaining the success of one paradigm over another. This shift in Kuhn's interpretation suggests that there are good objective, logical reasons for scientists proceeding as they do. The alternative explanation is that we term these reasons good largely because they are endorsed and rationalized by members of a certain scientific subcommunity—and that the real reasons for the substitution of one particular paradigm (among many competitors) over the previously reigning paradigm involve persuasiveness within the scientific community.

Indeed, we would argue that within science there exists a competition of theories and the mechanisms for their rational critique. The cultural imperatives of science, as well as the social structure of scientific institutions, promote not merely generalized skepticism but provide incentives for the systematic and rational critique of explanations already advanced. Paradigms, in the last analysis, are not exempt from this critique, even though the specific process of the critical examination of theories deals with the formally articulated and crystallized aspects of scientific thought, involving its paradigmatic phase more indirectly. It is in this sense that the replacement of scientific theory by better theory has been the trend of history.

If we apply the notion of paradigm in a general fashion, not restricting it to the scientific disciplines as such, we can see it as a significant ordering principle or structuring idea, not based on rules but on exemplars, that provides theoretical coherence and consistency in a circumscribed cultural domain of inquiry or body of knowledge. Consider, for example, the modern medical profession: significant professional subcommunities within core medical specialties—such as family practice, pediatrics, geriatrics, to name only the most obvious—subscribe to a model or structuring image of man or, human nature that rests on what is essentially a neo-Freudian, psychoanalytic, hydraulic personality paradigm. Similarly, one can point to significant literary and theatrical subcommunities composed of artists whose work is also inspired by such a paradigm. On the other hand, it seems quite clear that Kuhn's image is directly derived from fields in which there has been consensus on a single paradigm, where one can identify entire communities organized around a reigning image or model. Most sciences as well as most arts (or humanistic disciplines and professions) lack a single overarching paradigm. They are multiple paradigm fields (Ritzer, 1975). It may be more useful to think of an interplay between searches for new ordering principles and images (or paradigms) and minor additions in systematic but continuous work. Gerald Holton (1973) has suggested how lines of discovery in certain fields escalate because of seminal discoveries that lead to new directions of inquiry, while efforts simultaneously continue along the old lines of investigation. Holton's imagery suggests a branching of activity leading to escalation in the growth of research. As the pursuit of such branching lines of inquiry continues, the established boundaries of the disciplines become permeable and sometimes new interdisciplinary fields emerge or new disciplines are founded. This conception helps account for the increasing specialization and differentiation in knowledge production.

We have been focusing on science and scientific inquiry as if it were the same thing as knowledge production. At a certain level this is quite appropriate. Inquiry produces new insights. However, we are using the phrase knowledge production to refer to any systematic, organized inquiry. Obviously, significant problems of inquiry often emerge in a diffuse social matrix through debate, collective concerns, or the forming of Foucault's "positivities." However, the establishment and subsequent institutionalization of formally organized machinery designed for the systematic production of new knowledge require special attention. We turn now to this phenomenon in the professionalization of epistemic communities.

THE PROFESSIONALIZATION OF EPISTEMIC COMMUNITIES: THE DISCIPLINES

As an amateur enterprise, early science discouraged specialization as well as control over the formal qualifications of members of the epistemic community. The next critical stage in the professionalization of knowledge came with the establishment of the new scientific knowledge in universities and the formalization of professional scientific societies. Significant steps in this direction were taken in Germany between 1810 and 1830. During this phase, formal scientific communities were increasingly organized into disciplines for which there were formal criteria of membership and control over publications. After this time, these epistemic communities were no longer coterminous with cultural free spaces. In short, the professionalization of epistemic communities, concomitant with their organization into disciplines within the university structure, removed at least some of the characteristics of cultural free spaces from them.

The professionalization of epistemic communities in science started with the introduction of research-teaching laboratories and seminars, especially in the growing German university system between the 1830s and 1860s. Teaching and research were combined in the role of the professor who lived on his teaching salary and devoted part of his time to research. Research and teaching activities came to be located in the university and were given capital support. Advanced students learned their discipline through an apprentice system. An original research contribution became the qualification for admission to the epistemic community as well as a requisite for professional certification as university teacher.

The formative years of the American university system began in the late nineteenth century. Although some universities attempted to imitate the German example completely, American universities introduced a number of important innovations into the academic tradition and the structure of disciplines. Departments made up of several professionally qualified and at least formally equal colleagues replaced the single professorial chair with a research institute totally under the professor's control. The department replaced the chair as the unit representing a discipline. This departmental organization made for relatively autonomous, flexible units that permitted effective rearrangement of specialties and diverse disciplinary ideologies and loyalties. It proved to be far more flexible than the older German structure. More importantly, the departmental arrangement led to the regularization of research careers and the crystallizing of the role status of salaried professors as teachers and researchers. The group of formally equal professors gathered into a single university department, which claimed both to represent and to make original contributions to a specific discipline, constituted a new organizational form for epistemic communities. The regularization of the professorial role status and of orderly research careers determined by the discipline's specific criteria for valid methods of obtaining knowledge was of vast significance. It was one of the major bases for the emergence of an institutionalized knowledge production system.

The consequences of these developments were far-ranging. Members of departmental communities could undertake organized research efforts without facing rigid and stultifying internal hierarchies. The increased professionalism of department members engendered a strengthening of disciplinary loyalties that, in conjunction with increased job mobility after World War II, led academics to identify less with their university and more with their profession or discipline. The decentralized departmental organization within the American university paralleled the openly competitive and decentralized system of higher education that characterized relations between universities. This competition and decentralization led universities into an entrepreneurial pattern of mobilizing resources and seeking or creating markets for their services. This in turn facilitated the advance of knowledge developed within the university.

The professionalization of epistemic communities that resulted from the rise of the university department as an organizational form encouraged interaction and communication among members of the same discipline and strengthened disciplinary autonomy by clearly delineating boundaries in terms of professional credentials, qualifications, and jurisdictions. Thus, academic disciplines generated personal and professional networks for exchange of information and research collaboration.

The academic discipline is a special kind of epistemic community. It is typically organized around a delimited domain of inquiry that is characterized by a specialized frame of reference oriented toward specific abstract objects of investigation. The central focus of disciplines, whether physics or sociology, is the cumulative construction of an abstract body of knowledge about a specific class of objects that, in turn, delimit and define the domain of inquiry. The cumulative, incremental development of a body of methods follows from this focus, as does the enormous effort extended in the standardization of observation and reporting as well as in forming theoretical propositions. In a sense, academic disciplines are tradition-forming as well as perpetuating structures, a characteristic crucially necessary for the systematic accumulation of knowledge.

Academic disciplines also serve as major professional socializing contexts. They provide common filtered experiences for student cohorts in training that become the ground for personal and intellectual loyalties, as well as for an internal authority and reward system that is quite divergent from that in existence outside the disciplines. The ethos of academic disciplines emphasizes collegiality and places a certain value on equalitarianism. Within disciplines, however, differences in prestige are closely linked to differential productivity, which creates a powerful incentive and reservoir of rewards for generating extraordinary efforts and expending enormous energies.

In spite of their emphasis on rationality and systematic knowledge production, academic disciplines are capable of generating highly intensive motivations and competitiveness. Few contemporary institutions have insistently and successfully demanded such uncompromising loyalties and continuous efforts from their members as academic disciplines. The specialized selective recruitment and socialization mechanisms responsible for the far-reaching, sometimes even total, identity transformation that an academic, scholarly career involves are as yet insufficiently understood. However, it is clear that academic rituals of selection, induction, and graduation; the academic emphasis on intense competition, the repeated references by leaders of disciplines to heroic aspirations, landmark events, and inspiring achievements, all enter into the academic socialization process that takes place within specialized disciplines. The very effectiveness of the disciplines in absorbing motivational commitments from their practitioners may well involve their capacity to combine systematic, often repetitive, work requirements with a sense of the dramatic, of being oriented toward the forefront of knowledge. Even though most of the students entering any discipline will spend their lives as conscientious journeymen workers, or "working stiffs," they do tend to be oriented toward pervasive heroic images of a discipline's founding fathers and its giants of scholarship that direct the aspirations of students and neophytes.

Such myths and images profoundly influence members of the discipline in subtle, but compelling ways. They tend to legitimate the existence of an authority and reward system based on prestige and accomplishment. They contain models for what it means to be a serious member of the discipline and a worthwhile legitimate colleague. The availability of such role models, indeed conceptions of identity, is of the most profound significance for discipline-oriented identity transformations, both during and after specialized education and training. These conceptions symbolically structure the sense of work-related guilt and inadequacy that typically represents one highly effective consequence of disciplinary socialization processes and that distinguishes them from more conventional professional educational processes.

In brief, the motivational forces marshalled by disciplinary structures and processes are difficult to overestimate. We believe that they have contributed to the emergence of a new *estate* of science and scholarship, which is dedicated to continuous innovation and knowledge production, but which is also structurally organized into persisting hierarchies that are quite at variance with the otherwise secular trend towards equalitarianism and democratization.

There is no question that the universities long have tended to support movements towards greater equalitarianism and democratization in the society at large. This historical tendency derives from the liberal-humanitarian ethos of the university as well as its historical commitment to universalistic criteria for evaluating people and performances. However, this universalistic orientation tends to be overshadowed by an historical commitment to achievement. Especially, this involves an aspiration to excel in scholarship. This commitment to achieved excellence, combined with unrestrained competitiveness and rewarded primarily through the allocation of prestige, means that academic disciplines are among the most hierarchically structured and stratified epistemic communities in contemporary society. Thus, alongside the existence of egalitarianism in academic collegiality, in which lateral authority prevails among equals, there also is a more important vertical prestige hierarchy based on the achievement of excellence, involving significant innovations in theory, method, or application. As a consequence, an elite orientation permeates academic disciplines and endeavors. The balance between values of collegial egalitarianism, which has its roots in the norms of knowing, and values involving emphasis on the recognition of achievement may shift, but the dialectic tension between them will undoubtedly remain.

Highly specialized disciplinary contexts are generally sufficiently protected from outside pressures so that the intellectual
dynamics internal to the discipline can work themselves out. This
combination of the relative insulation against external pressures and
influences with strong incentives for continuous innovation as a core
internal dynamism accounts for the selective knowledge growth
capabilities and achievement of the academic disciplines.

We already have spoken of the importance of the department
as an element in university structure, supportive of disciplinary
autonomy and identity. A related structural innovation that contributed to the capability of academic disciplines involved the development of graduate and professional schools. The idea of a graduate
school, composed of departments organized along the lines of major
disciplines, reflected the acceptance of the role of the university as a
research-based teaching institution. In the United States, again in
contrast to the traditional German pattern, only the graduate student
was expected to become a specialist in knowledge production and
prepare for a research career. Students not interested in research
careers could limit their education to the traditional undergraduate
college or enter some professional school. This meant that graduate
schools could concentrate on the intensive training and socialization
of specialized, technical researchers.

American professional schools came to specialize in training
students for the practical professions, especially those with a scientific foundation. Professional schools in America were powerfully
influenced by the tradition of thorough practical training and applied
experience. The conception of scientific research in these schools,
which included applied, problem-oriented research, was highly
compatible with this practical orientation. In this sense, then, both
the professional school and the graduate school of arts and sciences
in the American university represented institutions training students
for independent professional practice or research careers.

By the beginning of the twentieth century, American universities were stimulating and supporting research on a scale that far ex-
ceeded the needs of training students and that was distinct from teaching. This institutional differentiation of the university and higher
education generally into three sections—the undergraduate college,
graduate school, professional school—and the provision for research
activity and interdisciplinary research institutes, only loosely tied to

teaching, opened limitless possibilities for the establishment of new disciplines and of innovations within disciplines. Two other factors that facilitated scholarly or research innovation within academic disciplines in the United States were decentralization and the resulting competition among universities. The absence of a centralized university system with monopolistic advantages meant that universities were forced to compete among themselves and to prove they were useful and worthy of support. This further strengthened the receptiveness to innovation of the American university.

The value of academic freedom has become probably the most central source of guidance and legitimation for the university and the disciplines and professions anchored in it. Indeed, the fight for the autonomy of universities from outside regulation took, early in their history in Europe, sometimes extreme forms involving physical struggle or the movement of entire faculties and student bodies to escape oppression. The modern conception of autonomy of the university, a conception closely allied to the idea of academic freedom, derives from this long history. It is probably the most explicit recognition of the importance of free spaces. European universities began during the middle ages as self-constituted communities of scholars, teachers, and learners. Most of these institutions were founded under the sponsorship of the church and existed to some degree under its authority. Prior to the mid-seventeenth century, the Roman church or its Protestant successors sporadically exerted controls and constraints on the universities or on members of their faculties, meeting occasional resistance. Even much later considerable censorship prevailed and limited the possibilities of the genuinely disinterested pursuit of truth.

It was in the eighteenth and nineteenth centuries that the political state became the sponsoring authority for most universities throughout the world, even though some remained under religious auspices. In the nineteenth century, the modern conception of the autonomous university developed in Germany and the idea of academic freedom was explicitly formulated in the Humboldtian ideal of the university. It was there that the idea of the autonomous university as a place in which scholars are to pursue truth, as well as to formulate and transmit it to students, came to be dominant.

In 1902, Professor Paulsen of the University of Berlin formulated a conception of academic freedom that had become accepted in his country during the preceeding decades:

It is no longer, as formerly, the function of the university teacher to hand down a body of truth established by authorities, but to search after scientific knowledge by investigation, and to teach his hearers to do the same. . . . for the academic teacher and his hearers there can be no prescribed and no proscribed thoughts. There is only one rule for instruction: to justify the truth of one's teaching by reason of the facts.

(Fuchs, 1964)

Paulsen also observed than any people who established and maintained a university,

cannot as such have an interest in the preservation of false conceptions. Its ability to live depends in no small measure upon its doing that which is necessary from a proper knowledge of actual conditions. And hence the people and the state . . . can have no desire to place obstacles in the way of an honest search for truth in the field of politics and social science, either by forbidding or favoring certain views.

(Ibid.:6)

This conception justifies academic freedom as a straightforward and direct consequence of the importance of truth. Matters are, in fact, a little more complex than that. The ideal of academic freedom in part derives from the need of knowledge producers and transmitters to enjoy the protection of a free space, in part from the autonomy demands of universities as corporate entities, and in part from pedagogical ideals. Thus, the conception of academic freedom in American universities has been extended to emphasize the rights of individuals to operate freely within their own fields. Individuals in the university must be free not only from external forces, but also from internal factions that could distort the process of rational debate and discussion in the search for truth.

It is certainly true that academic freedom in research and publication and in the classroom is augmented by personal freedom from institutional censorship or discipline when the faculty member, regardless of the field of study, speaks or writes as a citizen. However, MacIver (1955) and others have strongly argued that academic and personal freedom must not be confused:

Academic freedom is a right claimed by the accredited educator, as a teacher and as investigator, to interpret his findings and to communicate his conclusions without being subjected to any interference, molestation, or penalization because these conclusions are unacceptable to some constituted authority within or beyond the institition.

In the United States, the members of academic epistemic communities have placed far more stress on academic freedom for individuals, and on the mechanism to guarantee it (tenure) than on other aspects. Thus, the American Association of University Professors (AAUP) was formed as early as 1915 by a group of prominent faculty members in leading institutions, largely because of concern among these professors over dismissals that had recently taken place in a number of colleges. The purposes of the association were broadly professional; the earliest noteworthy statement of the group was the 1915 Declaration by its Committee on Academic Freedom and Tenure. The AAUP established itself as a leading force in the institutional definition of academic freedom, and also as its guardian. The phenomenon parallels the role of professional associations in relation to other epistemic communities; the guiding value also becomes a basis for legitimating economic claims and thus is linked to a professional ideology.

THE PROFESSIONALIZATION OF EPISTEMIC COMMUNITIES: THE PROFESSIONS

In addition to the processes that transformed academic-scientific epistemic communities, the nineteenth century also marked the crystallization of what may be termed the traditional model of professional practice; it marked the initial stages of professionalization among certain strategic, occupational epistemic communities. This raises the issue of the relations between the scientific and humanistic academic disciplines, on the one hand, and the practicing professions, on the other.

Freidson (1970:74) has suggested a useful distinction between the scholarly and learned (and scientific) professions, which create and elaborate the formal knowledge of a civilization, and the practicing or consulting professions, which have the task of applying that knowledge to concrete problems in everyday life. Although this distinction is more tenable than others that have been proposed (e.g., Ben-David, 1964:249), it conceals some rather obvious historical differences in the development of the two domains. The scientific and humanistic disciplines became professionalized in university departments after a long period in which most workers in these fields were self-supporting amateurs. The professionalization of certain occupational service communities that began in the nineteenth century took place with the decline of the much older guild system. The conception of professional work that began to emerge in the United States after 1800 was typically entrepreneurial and individualistic. The professional was viewed as servicing clients in a one-to-one relationship on the basis of individual knowledge and skills, acquired through some specialized training and resulting in some formal certification. Much of the knowledge and the command of the professional worker was acquired in practice, through apprenticeship. It was and remains role-embedded, implicit knowledge. However, increasingly the professions have engaged in efforts to systematize, codify, and organize their bodies of knowledge, standards of practice, and epistemic criteria. By no means have all professions reached the level of what Foucault calls "scientificity." However, a movement in this direction is certainly occurring—with profound consequences for the structure of the professions and their authority in society. We will return to the discussion of professions in the context of knowledge application, since the conception of professionalism in its very core emphasizes models and standards for knowledge use in practice.

THE PROFESSIONALIZATION
OF RESEARCH
AND THE BUREAUCRATIZATION
OF KNOWLEDGE

Numerous developments rapidly transformed the role of the scientist. Specifically, by the beginning of the twentieth century, there emerged in America for the first time the conception of the professionally qualified, institutionally certified, specialized scientific researcher. Before about 1900, research as such was not considered a profession, in the sense that scientific achievements were considered precious and fortuitous events that expressed and reflected the talent of especially gifted people. Research activities were considered voluntary, nonpaid, avocational activities. There were no institutionalized roles and positions that were successive stages of an orderly occupational career. By the end of the nineteenth century, however, the inadequacies of this situation combined with changes in the organization of knowledge in universities—specifically the emergence of the departmental structure for graduate and professional schools—to produce the professionalization of knowledge production by creating the role status and institutional position of the scientific researcher. Now a group of highly specialized research workers could pursue their intellectual interests in the context of a stable, orderly career.

The professionalization of research that began in the American university was intimately connected to the mobility and cosmopolitanism of researchers. That is, the creation of a set of interrelated institutional positions, which comprised a research career, encouraged scientists and scholars to become less identified with their universities than with their disciplines. In another sense, this development meant that in knowledge production domains, professional research orientations, commitments, and identities became antithetical to organizational orientations, commitments, and identities (Wilensky and Ladinsky, 1967; J. Marx and Spray, 1972). An even more significant consequence of the institutionalization of research careers was the emergence of a professional community of scientists or scholars in each discipline in which one's standing was more important than one's standing in the hierarchy of the employing university. One manifestation of the rapid growth in importance of professional communities was the expansion in significance of scientific professional associations in the United States during the first decades of this century. In short, only in the early years of the twentieth century—and initially only in the United States—was there general recognition that there is no inherent contradiction between creative, innovative research and intellectual achievement, and the institutionalization of professional research careers.

By the thirties in some sections of the country, and by the end of World War II throughout America, the organized production of knowledge by professional research scientists represented a large scale enterprise that had widespread public legitimacy and support. More importantly, the professionalization of research careers, the emergence of scientific or research entrepreneurs and administrators, and the development of standardized procedures and strategies for staffing, equipping, funding, and operating different types of research activity made scientific research into a transferable operation that could flourish in diverse institutional sectors and specific organizational settings. Specifically, these interrelated developments led to a burgeoning of scientific and scholarly research in industry and in government.

ORGANIZED KNOWLEDGE PRODUCTION IN INDUSTRY AND GOVERNMENT

In the period since World War II, very large and increasing resources have been invested in deliberate efforts at knowledge production by both industry and government in the United States. The result has been the emergence of a vast sector of the economy and institutional structure of the society, the research and development system. The pattern of the research and development organization emerged with increasing clarity before World War II. However, the war and subsequent years led to a very rapid expansion both in resources and numbers of organizations involved in such work.

By the early 1960s, there existed a highly decentralized set of mechanisms for the support of the research and development community by the Federal government. The organizational basis of the research and development system was extremely heterogeneous and pluralistic, growing by means of a kind of decentralized incrementalism that merged political and technical decisions. The growing magnitude of the funds involved, the growing complexity of scientific technology, and the desire of government to retain some control in planning and direction setting generated what Galbraith (1967) called the "technostructure." It was against this background that international competition, particularly with the Soviet Union, led not only to vast increases in government funding but also to the capacity to absorb such funds for significant knowledge production work. In the 1960s, the linkages between the research and development system and the government began to change. An increasing emphasis on applied or programmatic research concerned with governmentally set national policy goals emerged. Research was increasingly expected to identify practical problems that affected national welfare. Additionally, concern with potentially negative consequences of research and development arose. In consequence, the setting of science policy became a most complex endeavor. Never before were governments expected not only to assume responsibility for knowledge production, but also to attempt to anticipate both intended and unintended consequences of research.

The enormous scope of the national research and development effort in the United States is well known. Defense-related R&D, the space program and, increasingly, social problems such as educational reform serve as examples. It is clear that both novel forms of organization and of management methods have emerged in these efforts. The very nature of knowledge production activity requires a level of autonomy at work that is certainly higher than that in non-knowledge-related occupations. Modes of coordinating professional scientists cannot simply rely on directives issued by a bureaucratic hierarchy. The constraints inhering in knowledge itself, as well as its differential accessibilities, become powerful constraints for planning and managerial coordination. We will offer only one brief illustration from one of the less known domains of such efforts.

In the mid-sixties, the government established a research and development program in order to accelerate the process of producing and applying scientific knowledge to education. It was felt, at the time, that there existed in education an extraordinarily large lag between the recognition of a problem and the application of relevant knowledge to solve it. It was argued that the creation of major concentrations of resources and talents in governmentally financed, but university-based, research and development centers would lead to the type of sustained effort that would produce cumulatively beneficial results. Indeed, the program to establish not only isolated research institutes, but a national network of such centers, had as its objective the creation of a new type of epistemic community.

Since research for American education had, even without the government's interest in the topic, been carried on in many universities, colleges, and other institutions, it was obvious that the organizations set up as a consequence of the government's effort would account only for a minor, though weighty, component of the total system. Several research and development centers, and other organizations known as regional laboratories, were set up and became important sources for the production of new knowledge in professional education. This governmental effort was clearly linked to a hoped-for educational reform through increased professionalization of practice. The intention was that the art of teaching would be significantly enhanced by providing an explicit, scientifically valid knowledge base for the teaching professions. Not only psychology, but such fields as linguistics, anthropology, organizational sociology, and the like were pressed into service.

There can be little doubt that the teaching profession remains, in spite of all this enterprise, based on experience and art. The transformation of the profession into an explicitly knowledge-based one would lead to highly visible and consequential changes in teaching roles. This, precisely, is the objective of the R&D enterprise—the creation of an explicit knowledge base. Similar efforts are underway in other professional arenas such as that of law enforcement. Little is known at this time about the actual effectiveness or the unintended consequences of such policies.

These examples show that the R&D complex in post-modern society is not limited to the classic domain of physical technology. On the contrary, a great expansion into what is sometimes referred to as social research and development is taking place. Even though these new applications of the R&D model of knowledge production to fields involving social issues are based on the hope of replicating their success with physical technology, experience is only now being gathered on the significant differences in social science-based efforts. Indeed, it is necessary to distinguish between quite different patterns in technological and social research and development, especially where, in the latter, the focus is on the increased codification of professional practice. It is in these arenas that considerable social organizational and cultural innovations can be expected to occur.

The United States and Western Europe have moved toward the post-modern pattern with entirely different knowledge production systems and organizational structures. In the Western European countries, the new functions and responsibilities that organized knowledge acquired were attached to existing national systems of higher education. The tendency was to establish university centers of pure science. Research aiming at the solution of practical problems took place in segregated, specialized research institutes, usually financed either by government or industry and responsible directly to their source of support. Development occurred in industrial contexts, but was rarely organized and systematic. In the age of World Wars, European governments sought to stimulate research and development structures by establishing their own applied research and development structures by establishing their own applied research organizations or subsidizing such institutions established and operated by trade associations.

The American trend has been quite different in that knowledge production has shifted from specialized institutions of higher education to large universities and research and development systems performing an increasingly wide range of functions. There also has been the development of multipurpose institutions in the organization of industrial and governmental research. In the pluralistic, decentralized, competitive organizational context of the United States, trial and error rapidly established the marked superiority of large, multipurpose research organizations. Such multipurpose organizations encourage innovation in basic research and are particularly well suited to applied or mission-oriented research, which is likely to be interdisciplinary in nature. Moreover, the shifting boundaries between applied and basic research meant that the establishment of specialized research institutes—as in Western Europe—often froze research resources in specific subjects that soon proved to be of limited utility.

The critical difference between the R&D systems on the two sides of the Atlantic was the lack of centralization and the resulting competition that knowledge production institutions faced in the American context. This meant that American institutions were forced to innovate and learn from experience. This innovating consequence of the competitive struggle for institutional survival was, and to a large extent remains, largely absent in Europe.

Since self-governing university corporations have rarely been able to exercise much initiative in European countries, knowledge production and research policy making has usually been a governmental responsibility in its entirety. As a result, policy has been made by groups far removed from those responsible for its execution and by groups with little ability or opportunity to evaluate the results. Ben-David notes (1968-69:28-29):

> Paradoxically, therefore, the nationalization of the university and scientific research system which was supposed to lead to more objective and better coordinated planning of higher education and research has, as a matter of fact, debilitated the capacity of the systems to learn from experience, because the centralized systems had no constitutive feedback mechanisms such as are given by situations where universities and research institutes are free to make innovations and compete with each other, and because there was no room in these systems for the development of executive and entrepreneurial roles specialized in academic and research affairs, not too removed from its day-to-day activities and yet also not completely absorbed into them.

The organizational pluralism and entrepreneurship in the American R&D system in industry and government has, as we noted, given rise to an administrative style that combines high degrees of reliance on decision making by those close to the problem with a variety of techniques for coordination and supervision. This style has sometimes been referred to as a *new Federalism*. That is, administrative philosophy is formulated in analogy to the Federal principles vesting sovereignty in the United States, reserving to the central government only those authorities distinctly necessary for them. Such decentralized and pluralistic policy making, however, is difficult. The reliance on the mechanism of peer review, cost benefit analysis, programmatic budgeting, institutional monitoring, and the like increasingly codify these procedures and incidentally, create a mountain of records.

These developments raise the problem of policies for knowledge production, not merely in the global sense of setting governmental priorities, but in the more refined sense of deliberate planning. This planning cannot take the form that emerged in the older, hierarchical bureaucracies. It must reckon with the inherent structures and internal dynamics of disciplinary science and the institutional imperatives of knowledge production. Novel organizational patterns and devices having significance beyond the domain of organized knowledge production are emerging in order to coordinate efforts across different organizations and cultural systems, to facilitate the temporary composition of task forces, and to accommodate the autonomies of experts with widely differing epistemologies and disciplinary orientations.

KNOWLEDGE PRODUCTION AND IDEOLOGY

In the increasingly differentiated knowledge-based society, the various autonomous domains of inquiry develop epistemic criteria that often are widely divergent and potentially conflicting. Yet in spite of this, there is a widely shared conception that science provides for a unitary standard of truth toward which all specialized work is oriented. Thus, many professions such as engineering, medicine, clinical psychology, and teaching aim toward such a standard of scientificity, the principles of which are believed to be universal although the specifics vary drastically. Careful observation reveals that the specific standards embodied in the working conceptions of epistemic criteria vary widely. This is even more apparent if one compares one institutional domain with another. Certainly, the rules of evidence for clinical psychological inquiry and practice and those used in epidemiology, social work, or engineering research and practice are not identical or even roughly similar. Nevertheless, the frames of reference, reality tests, and rules of evidence for these divergent fields are reasonably systematic, describable, and embedded in a broader scientific conception of truth that does enable them to be translated into each other.

For the broad domains of technology, research and development, social planning, and that loose agglomeration we shall refer to as the helping-healing-human service professions, a broad conception of science serves as an ideal standard of truth. At the same time there are, of course, epistemic communities, such as those in the domain of the law, that have crystallized around quite different rules of evidence. Not that these domains are antiscientific, but they orient themselves to conceptions of the moral order using the specifics of sworn testimony and cross-examination as evidence, so as to acquire a separate status.

Yet the broadly defined conception of science is beyond doubt the strategically central factor in the knowledge system of post-modern

societies. This is so at least in part because the recent institutionalization of science as an organized, systematically planned and guided enterprise has repeatedly confirmed its capacity to formulate general laws
that transcend the reference frame of particular inquirers. Scientific
inquiry has shown itself capable of yielding coherently structured
knowledge that transcends the specific sociohistorical circumstances in
which it was produced. As a consequence, science occupies a privi-
ledged position in the knowledge system.

Foucault takes the view that science, in spite of all this, is
clearly ideological because of the very embeddedness of its pursuits
in the encompassing social structure. His argument is that certain
limitations following from this embeddedness simply cannot be
erased. By contrast, Theodore Geiger (1953) insists that there is a
sharp disjunction between ideology and scientific theory; he claims
that ideology is epistemologically false. For Geiger, ideological error
emerges when a thinker insists on smuggling subjective components
into his statements; by contrast, genuine theory is purely oriented to
objects and is thus free of subjectivity.

We find both of these seemingly diametrically opposed positions inaccurate. Certainly, Foucault is correct in emphasizing that
certain unreflected assumptions enter into scientific discourse as resi-
dues of the social structural terms and conditions defining a "positivity." Thus, it would not be difficult to demonstrate that many common (and common sense) assumptions about madness persisted for a
long time after the emergence of scientific psychiatry or that many
scientifically untenable assumptions about mental disease and mental
illness persist today within scientific psychiatry. At the same time,
Geiger's epistemological critique of ideology points in the right direction; ideological thought systems differ from those that are scientific
not necessarily in their content, but in the nature of the reality tests
they are subjected to and must survive. These, in turn, are a direct
consequence of the purposes for which they were constructed.

Geertz (1964) has argued that ideologies are cultural symbol
systems that arise in the absence of other resources for ameliorating
sociopsychological strains generated by the need for purposeful action in problematic or incomprehensible situations. More briefly,
ideologies represent shared cultural meanings that enable purposeful
social action in the face of uncertainty. We follow this definition and
emphasize particularly the idea that ideologies represent belief sys-

tems constructed around the need to present effective claims. In this sense, ideologies certainly can be found even within the most science-oriented organizations and groups.

The phrase *the end of ideology* was used and much debated as an assertion that political ideas had been exhausted and that there was a decline of intense commitment to political doctrine in the west during the 1950s (Bell, 1961). It remains a reasonably tolerable phrase if exclusively applied to political ideology in western, particularly American, contexts. The resurgence of political commitment in the early 1960s, on issues such as civil rights and peace, has not been accompanied by the development of formal political ideologies in the traditional sense. The decline of commitment to primarily political ideologies, however, does not mean the end of ideology generally or imply that the concept of ideology no longer possesses any heuristic value for sociological research. On the contrary, as Bendix notes (1964:295): "We may be witnessing a change in the arena of ideological conflict, rather than an 'end of ideology,' even in the Western context." From another perspective, "we are witnessing the 'end of some ideologies' rather than the 'end of ideology'" (Ibid.:324). We suggest that the main context and referents of Western ideologies have shifted from broad political issues and controversies to more highly differentiated and functionally specialized occupational collectivities.

CONCLUSION

This concludes our overview of knowledge production. We began with a consideration of the growth of knowledge, from which we turned to the study of social contexts and the forms of embeddedness of inquiry. These reflections gave us the background necessary for an understanding of the growth of autonomy in science and the structure and consequences of scientific norms for the dynamics of knowledge growth. Social structural considerations occupied us again when we turned to the question of the professionalization of epistemic communities, both in the disciplines and in the professions. Finally, we dealt with the questions of organized knowledge production as an industrial and governmental activity and its relation to ideology. Many of the themes struck in this chapter are likely to re-emerge in our treatment of other components of the knowledge system. We now turn to the questions of the organization, storage and distribution of knowledge.

7

The Organization, Distribution, And Storage of Knowledge

In this chapter, we turn away from the point of original knowledge creation to the manifold social functions that are involved in its organization, distribution, and storage. This aspect of the knowledge system has probably grown more rapidly than that of the knowledge producers. The gigantic mass of knowledge and specific items of information embedded in it that is the characteristic of current, post-modern society and culture poses special problems for the storage and distribution capacities of the knowledge system.

LINKAGES AND CAPACITIES OF THE KNOWLEDGE SYSTEM AND KNOWLEDGE TRANSFORMATION

Bodies of knowledge, far from being universally held or accessible, are in fact socially distributed. Specialized knowledge is available to relatively few individuals, sometimes with sharply defined locations in the social structure. Knowledge is commonly thought of as being organized into structures or entities, bodies of knowledge. It hangs together in more or less integrated and bounded entities. There is the body of knowledge of Linnaeus' botany, of

Newtonian physics, of contemporary information science, of mathematical logic, of psychoanalytic psychiatry, and the like. The variety in the formal structure and content of bodies of knowledge is enormous, yet one can discern certain basic, structural principles.

Bodies of knowledge are symbolically articulated cultural entities. They hang together through a principle of relevance that defines the necessary or contingent relation among items of knowledge. Such principles of relevance are embedded in the structure of bodies of knowledge. However, they also form the pathways through which one might travel from one of its regions to another. In this manner, bodies of knowledge are perceived as cultural entities or objects as defined by Anderson and Moore (1969). Such cultural objects consist of symbolically defined, largely coherent assemblies of information and cognitive skill that can be transmitted through learning. We will shortly turn to the discussion of knowledge structuring in this sense.

Bodies of knowledge are structured through cognitive activity and stored for social use of some kind. We need to remind ourselves, again, that they are embedded in finite social domains. The social structure of knowledge production and the nature and boundaries of epistemic communities have a profound influence on the selection of principles of relevance, which in turn structure the body of knowledge. Such social arrangements are support structures for the body of knowledge. Their significance for the problem of knowledge application will be discussed in the following chapter, through the theme of structures of relevance.

We already have observed that all knowledge is in principle relational. The activities of structuring bodies of knowledge, distributing access to them, and storing them require therefore the preservation and maintenance of appropriate knowledge skills and frames of reference. One might think of bodies of knowledge not merely as cultural entities but also as configurations of social activity, relating frames of reference systematically to each other. Allen Imersheim emphasizes this point as follows:

Much knowledge, specialized knowledge in particular, may be held only by persons in certain roles. But that the person in this role has this knowledge may be more generally known without knowing the content of the knowledge. Which is to say that if one does not know something in question, one may know who to ask to find out. While this may seem a fairly simple assumption which can be taken for granted, it has important implications for trying to relate knowledge structures that come from different aspects of a culture. It may be the case, for example, that the structural relation between two "bodies" of knowledge can be specified entirely in terms of the occasions when the persons occupying the roles with exclusive access to that knowledge come together in a common activity. Thus, knowledge structures cannot ultimately be separated from activity patterns. A consideration of knowledge structures is necessarily a consideration of knowledge communities.* *(Imersheim, 1977)*

Clearly, the structure of bodies of knowledge and their distribution in a society is constrained by the nature of knowledge-related activities and the communities in which they are embedded. At the same time, the structure of bodies of knowledge becomes a significant constraint on knowledge communities and activities. Criteria of appropriate conduct for knowledge workers in epistemic communities not only fix the structure of bodies of knowledge as cultural entities, but are also limited and constrained by them.

It is useful to think of the knowledge system as providing a distribution of knowledge capacities and linkages among them. The system is analagous to a complex network of channels of communication through which knowledge flows. Much of the literature concerning the dissemination of innovations and knowledge has conceived of social structure as a web of pathways and gates through which knowledge flows at quite different speeds. (See, for example, Rogers and Shoemaker, 1971.)

Recent discussions of knowledge diffusion portray the distribution of knowledge from a center, the locus of its scientific production, to the periphery—presumably the domain of its use. An interesting alternative process has been examined by Dorothy Rosenberg in her study of the incorporation of acupuncture into scientific American medicine (Rosenberg, 1977). Knowledge flow processes do indeed occur in several directions through social structure; the focus on the center to periphery direction is only one among several possibilities. This chapter examines a number of fundamental issues relat-

*Allen W. Imersheim, "The Epistemological Bases of Social Order: Toward Ethnoparadigm Analysis," pp. 1-51 in D.R. Heise (ed.), *Sociological Methodology*. San Francisco: Jossey-Bass, 1977.

ing to knowledge organization, distribution, and storage, especially in post-modern society. Indeed, it is this domain of distribution and storage of knowledge that has particularly expanded in the transition to the post-modern stage. The preservation of knowledge, its guardianship, has been the function of well-established institutions in traditional society. The proliferation of specialized arrangements, organizations, and mechanisms for knowledge storage and distribution is a special phenomenon of post-modern society. Fritz Machlup (1962) estimated that about 29 percent of the GNP in 1958 was spent for knowledge. However, his definitions were extremely broad. He included not only education in the formal sense, but also education on the job, in the home, and in almost any other setting. His category referring to communications media included commercial printing, stationery, and office supplies. His definition of information services included, for example, any money spent for or by real estate agents, tax consultants, security brokers, and the like. In spite of this breadth of categories, his work is most instructive. He estimated about 72 percent of the total societal resources spent on knowledge in 1958 were devoted to knowledge distribution and storage functions. Since Machlup's study, this concentration of resources has doubtlessly become more pronounced.

As a consequence of the strategic importance of this aspect of the knowledge system, the post-modern pattern is characterized by the development of explicit cultural models mediating and connecting the search for knowledge with the utilization of knowledge. Such models have been institutionalized into specific role statuses and positions within organizations in a self-conscious attempt to promote a smooth flow of knowledge through the system. This chapter focuses on those symbolic models and structural arrangements that have been developed to organize specialized, technical knowledge into coherent bodies or packages; to distribute and disseminate information deriving from these bodies of technical knowledge to appropriate users; and to store the rapidly accumulating bodies of knowledge in such a manner that they are readily accessible and retrievable as the need arises.

Several qualifications are prerequisite to a discussion of cultural models and social structures that organize knowledge flow. First, there is an obvious overlap of aspects of organization, distribution, and storage in the real world, particularly since all aspects of the process involve communications media. For this reason, the issues will be illustrated through examples concerning communications among professional and scientific knowledge producers and users. One further focus will be the reporting of technical knowledge in the mass media, since significant knowledge transfers occur in this manner, especially into the cultural system of common sense. Knowledge storage can be seen more distinctly, since central to this social function is a clearly crystallized institutional model, the library. This traditional model itself is in a rapid process of change into technological forms of great diversity and scope.

We also need to clarify our use of the term *media*. This designation refers to communication and interaction channels—vehicles of symbolic meaning that facilitate the communication of specialized symbol systems. In our sense, then, scientific conferences, professional meetings, and work sessions represent media just as much as textbooks, journals, newspapers, or television shows. What they have in common is that, in a certain complex fashion, they influence the organization and structure of technical knowledge. Our use of the concept of media is most familiar with respect to those channels that distribute specialized, technical knowledge in a form that can be understood by lay publics. In this context the written and electronic mass communications media figure prominently.

When we address the subject of knowledge storage and retrieval systems, the concept of media is employed in an unfamiliar fashion. Here we are thinking primarily of the physical or mechanical device used as a carrier of symbolism and its immediate social context. Thus, the body of technical knowledge can be preserved in the media of stone tablets, books, data cards, magnetic tape, voice recordings, film, and others. Obviously, the physical object receives its significance through the potentialities for use. It is embedded in an immediate social context, in which the relevant knowledge retrieval skills—often highly specialized ones—stand out. Clearly the system of knowledge storage interacts with the distribution of knowledge retrieval skills so that one needs to learn how to use a library's card catalogue system, cross reference data archives, computer programs, and the like.

The enormous complexity of the problems touched on in this chapter precludes attempting comprehensiveness of coverage; rather, issues were selected on the basis of their significance as examples.

ON THE ORGANIZATION OF
BODIES OF KNOWLEDGE:
CULTURAL MODELS, PARADIGMS,
AND KNOWLEDGE AT HAND

From one perspective, the organization of cultural symbols into coherent packages of meaning, which we refer to as bodies of knowledge, always engages the cognitive structuring processes of the individual. Although bodies of knowledge are organized in relation to individual knowing, their design follows principles of symbolic coherence and articulation as well as social structure, which extend well beyond the knowing processes of any given individual. Bodies of knowledge are never presented to individual knowers in unstructured form. Instead, they exist in some culturally determined form, and are organized in terms of some cultural model.

The organization of bodies of knowledge occurs at least at three different levels. There are broad cultural models, more specific paradigms, and arrangements of information into bodies of knowledge for the convenience of use (knowledge at hand).

Cultural models and paradigms can be said to determine what information is considered to be the prerequisite of further information; which symbolic elements ought to be given central importance and which are peripheral; what is the focus of attention and what merely the contextual background; and what are the boundaries of the groups of symbols and what bodies of knowledge become an important influence on the location of problems and their definition. One can say that such symbolic structures are codeterminants of the production or discovery of knowledge as well as its organization, storage, and distribution.

At the level of broad cultural models, for example, are the subtle effects resulting from the idea of linear, temporal sequences that have played such a profound role in western civilization. In this cultural model, knowledge is organized according to the underlying principle of successive ordering in time. A close relationship exists between this linear conception of the progression of events through time and the idea of causal reasoning. In contrast, there is the cyclical, knowledge-organizing model in which recurrent, rather than linear, nonrecurrent time is the structuring principle. The conception of progress would be quite alien in a context in which such a cyclical organizing model was used.

What we have in mind is quite closely related to what Vytautas Kavolis calls "conceptions of order" (1975).

> Conceptions of order and disorder tend to be implicit and global and can be expected to be located on a deeper and in any case, less consciously understood (and more intuitive) level of the symbolic organization of civilizations than do the more explicit and differentiated cultural logics. . . . Four symbolic frameworks for comprehending the basic character of order have had particularly important implications for the organization of individual identities. They can be imagined as "lawful nature," "spontaneous nature," "the factory," and "work of art,"—as metaphors of order somewhat comparable to the Greek categories of *logos*, *physis*, *nomos*, and *eros*.

Kavolis relates these conceptions to the problem of individual identity; we find them useful in thinking about the fundamental structures of bodies of knowledge. He argues that basic metaphors, derived from some concrete experience of order, become extended and serve as broad ordering principles. In conceptions of nature, the experience derives from some primordial order existing independently of human intentions. In the conception of the "factory" and the "work of art" reference is to experience with human creations of order. It appears convincing to us that such basic metaphors of order or cultural models form the taken-for-granted context out of which more specific knowledge-ordering activities arise. Kavolis' usage points to the significance of exemplification and the extension of perceiving order and structure from examples.

This is a point that, following Kuhn, is treated by Allen Imersheim (1977). He discussed the role of "paradigms" (Kuhn, 1962, 1970, 1974) not only as sources of structure for scientific theories, but also in their form of "ethnoparadigms" as bases of social order. To him paradigms define ways of seeing. A given datum does not remain fixed in its meaning all the time, but may be very differently perceived depending on the observer's perspective and paradigm.

> If there is a difference between what scientists report, it is not because they interpret data differently, but that they see data literally, in a different way, perhaps even see different data. The shift in perception from one paradigm to another is like a visual

Gestalt switch. First one sees a duck, then one sees a rabbit. To be sure, for this change to occur other related elements in the "picture" have changed—the figure is viewed as related to other elements in a way that it was not before, it is seen in a new light. But, data are seen as data only by virtue of relying on a paradigm. Furthermore, Kuhn argues, one does not see a figure and then interpret it in the light of a theory to be a rabbit. Rather, one *sees* a rabbit!

(Imersheim, 1977:26)

Imersheim emphasizes the role of exemplars in defining these cognitive processes. The problem of exemplification in making meaning clear is a complex one, but it plays a major role in the structuring of bodies of knowledge and in knowledge transmission. Exemplification relates understandings to a context and an activity. As knowledge is transferred from one context to another, the learning of specific rules, symbol systems, and frames of reference helps; but exemplification is probably one of the central processes. Imersheim writes (1977:27-28):

... components of a paradigm, and perhaps they are ill-termed as components, *possess* their characteristics as components not by virtue of what they are, but by virtue of how they are used (or how they might be used in the case of a paradigm not yet adopted). Not just any example will do to be an exemplar; it must be a central example, one around which a wide range of activity is or may be organized. Not just any set of laws or symbolic generalizations will do, but they must be able to be used in a certain way; in particular, their meaning is given through their use in examples.

It is in this sense that paradigmatic exemplars as well as explicit modes of symbolic coherence come to be structuring principles for bodies of knowledge at a level more specific than the general cultural models we have discussed earlier. Systems of classification, such as Linnaeus' classification of plants, may serve as simple illustrations of this principle. More complex ones are of the theoretically paradigmatic kind, such as the Darwinian theory of evolution. Finally, we can point to a principle for knowledge ordering that derives from more pragmatic demands of activity. This is the assemblage of information packages so that they are readily at hand in situations of need. For example, the typical handbook for practice arranges knowledge not so much in terms of its inherent consistency and coherence, nor in terms of an overarching paradigm, but in terms of anticipated knowledge needs. Diagnostic clues of diseases are listed with possibilities for treatment. Assembling an abstract of traffic laws, mechanical information on automobiles, and safety rules is justified by the demands for practice in driver education.

The actual manifold possibilities of organizing knowledge, no doubt, are derived from a variety of such factors. Broad cultural models, the nature of symbolism itself, the structure of theories and arguments, exemplars, and pragmatic assemblies of information all are aspects of these structuring processes that shape knowledge patterns.

ON THE ORGANIZATION OF TECHNICAL KNOWLEDGE

There are many ways in which technical knowledge may be organized so that it becomes available to specialists. One of the most prominent patterns is the scientific or technical professional journal. One survey of American scientists found that they obtain their information primarily from "journals regularly scanned" (30.4 percent), from "citations in other papers" (10.9 percent), from "author reprints" (5.8 percent), from "abstracting or indexing services" (6.4 percent), from "compendia" (4.3 percent), from "casual conversation" (22.1 percent), "by asking colleagues" (8.1 percent), and through "other methods" (11.5 percent) (Glass and Norwood, 1959). In other words, nearly 58 percent of technical data comes through written media, most of the rest through direct interpersonal communication.

Bourne (1962) estimates that over a million scientific or technical papers are published each year in journals, bulletins, and reports. It is reported that this output is doubling every decade or so. Hagstrom (1965:23) agrees that "formal communication in the sciences is primarily carried on through articles appearing in scientific journals." In his analysis, formal communications represent a part of the system of recognition and reward that performs fundamental social control functions in science. Citations of others' formulations, of course, constitute acknowledgments of their significance.

The estimate of a million articles a year appearing in scientific journals is less overwhelming when one remembers that scientific articles and journals tend to be highly specialized and segmented. In consequence, researchers in a given area have a much more limited number of sources to review in order to keep up with their work. At the same time, the available evidence indicates an extremely high rate of growth of information even within highly specialized areas.

The Glass and Norwood (1959) survey also indicates that collaboration is a very intensive form of communication conducive to productivity in science. Each specialized area of science seems to contain a very active core of researchers, around whom there evolves a large "floating population" of researchers who collaborate with core members on specific team-authored papers. Although each of these subgroups generally centers on the institutional location of its core members, they are not necessarily so confined. The fundamental importance of formal, planned communication channels in science—such as journals, research collaboration, conferences—should not obscure the significance of informal and unplanned communications that occur outside institutionalized channels. Hagstrom (1965:31) has characterized the differences in formal and informal channels in science in a particularly clear way:

> Formal channels of communication demand responsibility: the scientific article is expected to be a finished and polished piece of work. Informal channels of communication, on the other hand, often involve a great deal of permissiveness. People can make suggestions without committing themselves, and others can criticize the work without having to make any final decision with regard to its validity.

Preliminary and untried ideas, information about experimental or organizational procedures, and the clarification of particularly difficult points are among the items apt to be transmitted through more informal channels of communication.

It is clear that the mode of distribution of knowledge through formal versus informal channels is interdependent with the context of knowledge use. The reliance on formal communication in basic science is readily comprehensible in this light. Similarly, one can understand the very different relation between formal and informal communication among practitioners. Applied scientists and engineers tend to rely relatively more heavily on informal channels and conversations with colleagues. In these circles the formal journal article plays a lesser role. In a comparison between seven professional engineering projects and two scientific research projects in physics, it was found that 51 percent of the ideas in the scientific research came from the formal scientific literature. However, only 8 percent of the ideas in the engineering project came from this source. On the other hand, in the engineering projects, respondents reported that 33 percent of the ideas came from vendors of potential equipment or subsystems or from the customers of the products being developed. These sources of information did not play any role in the research projects.

Differences are revealed in the structure of bodies of knowledge as well as in their use. Practicing professions almost inevitably make considerable use of role-embedded knowledge and pragmatic knowledge convergencies around a pattern of activity. The channels of communication and knowledge distribution reflect these patterns. It is interesting how profound the integration is between knowledge search behavior and the structure of bodies of knowledge. It appears to us that this complexity must be kept in mind when one thinks of knowledge diffusion, cautioning against the acceptance of any simplistic model.

As we have seen, applied scientists and professional practitioners have a greater dependence on informal channels of communication whereas scientific researchers are more exposed to an influence by formal channels, such as technical journals, monographs, and research reports. Because of the greater dependence on informal channels, patterns of communication among professional practitioners are more influenced by interpersonal factors and forms of association. Moreover, these informal communication and interaction systems among applied scientists and professional practitioners are highly influenced by their value systems. This theme recurs particularly in relation to knowledge use and the role of ideology in reducing uncertainties among professional practitioners.

Our discussion of the reliance of research scientists on increasingly specialized, technical, and formal means of communication, which are proliferating at an astounding rate, raises a central problem involved in the organization of knowledge. As the sources of information increase, individuals have increasing difficulties integrating the new information into their worlds and their work. Thus, Weinberg (1967:42) applies the idea of increasing semantic and conceptual complexity to the development of specialized fields of scientific inquiry. The world of science, he argues, has become incomprehensible in its larger contours because the increasing fragmentation of research has confined each scientist to a narrow field. As this specialization continues, fields of interest and activity become narrower and the number of people who can integrate related specialties declines. The immediate danger is loss of professional efficiency, but the long-term damage will be that science must further remove itself from the real world by advancing to ever higher levels of abstraction. As Weinberg explains (1967:43–44),

> Growth and fragmentation impair the efficiency of science by forcing science to become a team activity, because a single knowledgeable mind is in many ways a more efficient instrument than is a collection of minds that possess an equal total sum of relevant knowledge. . . . As science fragments, it seeks to reintegrate itself by moving to a higher level of abstraction.

Progressive abstraction becomes the only way in which science can reintegrate itself to cope with increasing specialization. Matters once grasped immediately, almost existentially, must now be grasped in principle. But knowing in principle is not the same as knowing from experience, for as ". . . one moves to a higher level of abstraction, one omits something, either because applying a complete theory to a detailed situation would go beyond our mathematics . . . or because our original theory omits something initially" (Ibid.:44–45). The price of increasing abstraction is paid by an increase in specialized terminology and esoteric conceptualization. Moreover, the unification of scientific knowledge at progressively higher levels of abstraction creates conditions that increase the scope of comprehension, as well as drastically increasing its distance from common sense. Fragmentation and abstraction are apparently inevitable as knowledge production increases. At the same time, one may argue that through abstractions new simplifications are made possible from which further knowledge may grow.

Garvey, Linn, and Nelson (1970) suggest several other interesting observations about specialized, technical communications associated with different scientific disciplines. Research that resulted in a published article took about a year for both physical and social scientists to complete. However, the former took an average of six months to prepare and submit a report of their work, while the latter averaged nine months between completion and submission. Whereas physical scientists experienced about a seven-month lapse between the submission and the publication of specialized, technical papers, the equivalent average for social science papers was nearly twelve months. Thus, the specialized, technical journal as a communications medium for organizing knowledge still remains a reasonably rapid and extremely versatile mechanism.

Despite considerably greater delays experienced by social scientists in bringing their work formally and in organized fashion to the attention of their colleagues, they were less likely than physical scientists to make use of other channels of communication. Thus, 83 percent of physical scientists disseminated their work to others in some way (usually through working papers) before submitting it for publication, compared to only 72 percent of the social scientists. Similarly, 75 percent of the physical scientists and 66 percent of the social scientists had made some report of their work to their colleagues through colloquia, working papers, or technical reports.

This seems to indicate that there may well be more pressure in the physical sciences to publish in order to document priorities of discovery than in the social sciences. Moreover, there is a higher rejection rate of articles experienced by social scientists that may well be evidence of a relative lack of consensus on standards of excellence in such fields. These observations illustrate differences in communication patterns and the structuring of bodies of knowledge across fields. Another source of influences on knowledge structure derives from preparing bodies of knowledge for pedagogical purposes in teaching. The textbook is the primary medium in this area. Its organization of knowledge may differ, sometimes drastically, from the organization of knowledge for research. The emphasis on coherence and comprehensibility in the textbook presses for knowledge integration and relative simplicity. The sequencing of topics often does not follow the inherent logic of research, but presents arrangements thought effective for instruction. Yet major textbooks have influenced the thinking in scholarly fields for long periods, establishing a common frame of knowledge and interpretation for the students of a discipline.

The frames of reference of scholarly fields have been discussed earlier; they established both the enabling grounds for knowledge production and communication as well as barriers for knowledge transfers. While conceptual models and sensitizing ideas may be transferred from the qualitative fields to quantitative ones, it is generally the case that specialized, technical knowledge tends to flow from the quantitative to the qualitative discipline. It is the realm of research methodology, design, and instrumentation that has proven most amenable to organized transfers of knowledge from one field or discipline to another, and the one in which communication takes place with relatively less difficulty than other knowledge transfers across reference frames. The list of cases in which a research technique orginating in one discipline has been picked up by workers in another field is endless. For example, statistics was first brought to a level of wide-ranging sophistication and utility by agricultural scientists who needed it to help in interpreting their experiments in plant breeding and nutrition. At the present time, it is associated heavily with the social sciences. The ease with which methodological knowledge can be organized in such a fashion that it can be transferred to other fields is probably a result of the fact that the data-generating and interpreting advantages of a new research technique are frequently clear-cut, whereas the theoretical utility of a concept borrowed from another field may be a matter of dispute. But even in the field of methodology transfers, boundaries remain significant, a phenomenon that is much influenced by the qualitative modes of coherence in the body of knowledge that we discussed in relation to cultural models, frames of reference, and paradigms.

THE POPULARIZATION OF KNOWLEDGE AND COMMON SENSE

Knowledge distributions and transfers, we have seen, are massive in scope and involve large social resources. Mass education, as reflected by the expansion of schools, colleges, and universities throughout the world, is one major aspect of the phenomenon. The increasing reliance of professional expertise on formal, codified, and therefore replicable and also learnable knowledge in lieu of exclusive reliance on some implicit art (such as the art of medicine) contributes to the process. Citizens' initiatives in controversies concerning environmental protection, health, or other problems can and do mobilize explicit knowledge resources and begin to check the authority of the professionals. Such mobilizations accomplish significant knowledge transfers. The possibility of this phenomenon rests on the availability of knowledge about knowledge, that is, of information about knowledge resources and their accessability. It is in this regard that the mass media play a major role.

The exploration of space has become an exemplar of science for the public. For example, the Pioneer X mission to Jupiter in 1973 was such an event. *The New York Times* carried the following editorial on December 4 of that year: "Now the data from Pioneer X inaugurate a qualitatively new stage in man's knowledge of this giant planet and in his understanding of how its physical and other properties relate to the larger problem of the origin of the solar system and of the universe." On the next day, the *Los Angeles Times* wrote: "Pioneer X is a triumph of technology, but it is more than that, it is another successful attempt by endlessly curious man to puzzle out the secrets of the endless universe and then, someday, turn them to his own use." The long voyage of Pioneer X to the planet Jupiter climaxed on December 3 with a precisely executed fly-by 81,000 miles above the planet's clouds. The Pioneer Mission offered the public an opportunity to witness glimpses of science in action. As the space craft continued to radio data back to the Mission's operation center, the principal investigators engaged in daily interviews with reporters. Dozens of reporters, many of whom were specialized science writers from major newspapers, magazines, and wire services attended every briefing by the scientific teams. These daily briefings began a week before the rendezvous with Jupiter and continued each day until a week after the encounter. Some of the discussions were relatively brief progress reports on the functioning of Pioneer's instruments; others lasted several hours and offered detailed interpretations of the mission and the questions it sought to address. During these briefings, reporters pressed for details on the meaning of the mission and the data Pioneer was sending back. The scientists, elated at the consistent success and perfect functioning of the instruments and the space craft, were eager to answer at length.

In one sense, for the public, this represented instant science. It was different from watching the spectacular feats of manned space explorations and voyages to the moon. A sense of excitement and discovery, of questioning and of learning, continued to surround the Pioneer news coverage and generate sufficient interest to command front-page attention for several days. Newspapers all over the country featured the Pioneer story prominently. Even writers who were not personally sent to California to cover the story wrote interpretive commentaries. Television networks carried news reports on the mission, with interviews, charts, and models to explain the data.

Occasions when a scientific event captures the public's imagination and successfully competes with entertainment television are rare. As Etzioni and Nunn (1974) have empirically demonstrated, public confidence in the institutions of science has declined in recent years. However, public confidence is very different from public interest and interest on the part of the public in science may, in fact, be growing. This level of interest and attention is nourished by many sources.

Both of the major American news magazines, *Time* and *Newsweek*, devote significant space to their sections on science, medicine, and behavior. National newspapers frequently publish long articles on scientific issues. Professional meetings of the national associations of the natural and social sciences receive widespread coverage in the press. Reports in scientific journals that are considered interesting by reporters are regular sources of news and frequently are accompanied by extended personal interviews with the investigators. These activities, which are geared implicitly to the transformation of common sense, have generated a number of new interstitial role statuses that mediate between knowledge producers, the general public, and knowledge users.

One such role status that has considerable importance for mediating between scientists and various publics is performed by science journalists. Until the 1930s, science writing for media was virtually an unknown specialty in journalism. The National Association of Science Writers was founded in the bar of a Philadelphia hotel during the annual meeting of the American Philosophical Society in 1934. A more formal meeting in Washington later that year, at the annual meeting of the National Academy of Sciences, launched the organization. The first eleven active members included virtually all the full-time science writers in the United States in 1934. By 1938, the National Association of Science Writers had grown to 18 members; by 1945 NASW could boast of 61 members; the number of members reached 200 by 1963; and in 1974, there were about 400 full members of the NASW. Note that the membership increased a little more than three times between 1938 and 1945, a little over three times between 1945 and 1963, and despite the fact that a larger base is involved, doubled in the ten years between 1963 and 1974. Although entry into the field of science journalism remains unsystematic at best—in that very few science journalists have had specialized, advanced training in the sciences—most science journalists have had considerable experience in more general newspaper reporting before they specialized in science. Therefore, their core identification may be with journalism rather than with the scientific field on which they are reporting. However, this group of specialized science journalists exercises an influence on public views of science that is far out of proportion to their number. And it is they who, as the example of Pioneer X shows, have exerted great influence on the development of post-modern common sense among the general public.

Other such interstitial roles have emerged in many institutional domains. For example, government agencies in many technical fields have become aware of the need for science *translation* services. Technical data about the properties of complex technologies, demographic trends, threats to health and safety, and the like need to be transformed into a readily understandable version for policy makers and the public.

Such work is especially important in the preparation of educational materials. Curriculum design has become a virtual industry, not just the activity of individual textbook authors. So has the preparation of instructional films. The role of educational television, correspondence courses, and continuing adult education can hardly be overestimated in the impact on the distribution of knowledge.

More specifically, the relation between knowledge-producing organizations and their potential clients or customers has often been structured through specifically devised linking mechanisms. Sales representatives in high technology fields are often much more than salesmen; namely, agents of knowledge transfer. Similarly, relations between research and development branches of industrial firms and their management and manufacturing components seem to require roles of this kind. It is difficult to say where the general function of knowledge popularization leaves off and the specific function of bridging frames of reference and aiding knowledge transfers begins. It does appear to be the case that many instances of knowledge use are mediated through general rather than specific linkages. Certainly with regard to the impact of social science work on policy, the argument may be made that policy makers rely less frequently on specific studies submitted to them and more frequently on broader channels of communication and knowledge made available in the context of public debates.

KNOWLEDGE STORAGE AND RETRIEVAL

The subject of knowledge storage and retrieval is enormous; we can only illustrate it. Its significance lies in the fact that the scope of knowledge socially available far transcends the capacities of all individual memories. Social and cultural mechanisms for knowledge storage and retrieval in fact constitute a societal memory. A great deal of knowledge is stored implicitly in traditions and in the role expectations built into social structure. But there is a world of difference between reliance on living memory and explicit, codified, systematic knowledge storage with provisions for easy access and retrieval. To emphasize the already obvious point, one only needs to contrast the knowledge storage capacity of a traditional peasant village with that of a modern university with its library system, computer-based data banks, film collection, reference services, collection of abstracts and more. Since knowledge storage in memory and easy retrieval in a situation of need are key ingredients for learning, one might hope that the vastly expanded institutional capacities for knowledge storage and retrieval would lead to improvements in the ability of collectivities to learn from their own past experience and improve their future performance.

The enormous institutional innovations in the last century, which have resulted in new mechanisms and structures for the production of knowledge, have been accompanied by equally important institutional arrangements and technical devices for storing knowledge. Libraries and archives have been in existence for a long time; however, the establishment of large administrative units and professional services for the purpose of not only gathering, organizing, and preserving knowledge, but also for storing it in a readily retrieved form, is a relatively recent development.

Record keeping, accounting systems, and rational knowledge accumulation for the purposes of economic activity and policy were central features of the rise of modernity. Commercial corporations, state governments, and other administrative bodies contributed as much to this development as did the sciences. Indeed, one might argue that there is a dialectic inherent in the expansion of bureaucracies. They are, and usually have been, viewed almost only as instruments of control. At the same time, however, they are mechanisms for social measurement and the storage and retrieval of social knowledge. It is difficult to imagine the rise of social self-consciousness and, indeed, the development of revolutionary thought and movements without this enormous, expanding base of social knowledge. Steensgaard (1976) has demonstrated that the great overseas trading companies of the seventeenth century owed much of their success to the development of a rational accounting, reporting, and intelligence system. Within the extremely brief span of two decades, these companies deflected the trade from the Far East and Europe from traditional caravan routes to overseas maritime transportation channels that they controlled. Through the establishment of an organizational memory and intelligence system, they were able to reduce the level of uncertainties and derive commercially and politically successful policies.

In recent decades, far more specialized structures for knowledge storage and retrieval have arisen, specifically in the form of modern library systems, museums, and bibliographic or information services. Libraries represent basic knowledge availability systems that are far more than mere repositories for storing books. Changing library designs over the past hundred years have reflected and been closely associated with changing conceptions of the underlying rationality and order in knowledge. But problems of knowledge storage and availability have become so massive in recent decades that there has been a proliferation of disciplines, professions, organizations, and services addressed to the task of knowledge storage and retrieval. Thus, computer-based data banks and information systems are increasingly supplementing libraries; such electronic information systems may even ultimately supersede libraries as they are currently known.

These kinds of developments have led to the emergence of new professions and disciplines whose main intellectual and practical responsibility is for the management, storage, and retrieval of bodies of knowledge in a formal, rather than a substantive way. Moreover, emerging professions, such as that of information science—superseding older conceptions of librarianship and library science—inevitably incorporate a degree of reflectivity about knowledge unheard of in past structures for information storage and retrieval. Instead of the earlier predominance of a substantive focus on the classification and storing of relevant bodies of knowledge, these new disciplines, domains, and techniques focus on structures of relevance, ways in which information can be traced within bodies of knowledge, and ways of charting the various channels of knowledge flow through social systems. In other words, knowledge storage systems in postmodern society have evolved to such a high degree of complexity that their understanding and management require specialized expertise.

There is one particular respect in which recent electronic innovations relevant to the storage and retrieval of knowledge bear on some of our earlier points concerning the professions. Specifically, the use of computer storage and retrieval capabilities for accessing the more formal, academic component of professional expertise has profound implications for the future activities of the established professions and their reliance on a body of specialized knowledge.

Wilensky (1964), for example, has argued specifically that professions differ from other occupations in that they are based on a body of systematic, esoteric knowledge acquired through long training in the course of which incipient practitioners come to internalize professional norms involving a service ideal. High status, established professions have gone through a long process locating their training within universities and establishing formal professional associations that press for licensing and certification procedures and define ethical codes. The main consequence of these processes has been that the established professions have an extraordinary degree of autonomy. In this view the critical characteristics of a profession center on its body of specialized, systematic knowledge, a service ethic or ideal, and high degrees of authority and freedom resulting from work autonomy.

There is, however, an important qualification to this characterization; it is not knowledge per se that produces autonomy but, rather, the monopolization of a body of specialized, systematic knowledge. As Freidson (1973:28) was quick to point out, "knowledge itself does not give special power: only *exclusive* knowledge gives power to its possessors." Moreover, it is this monopolization of exclusive knowledge that the professional possesses over a particular area that generates the charismatic aura of mystery and the myths about particular virtuoso professional performances. Such exclusiveness is easier to maintain with regard to role-embedded, implicit knowledge and skill, the art that is not readily transmittable. There is a paradoxical situation in that rising levels of education can result in greater utilization of professional services at the same time that they also produce a greater sophistication about the assessment of these matters, and a desire to check on professional performances by comparing them to standards derived from codified, explicit, and formal knowledge. Under these circumstances, publics are likely to be more critical and less deferential than under other conditions.

This situation may well be exacerbated by the introduction of computer storage and retrieval systems into professional training and practice. We suggest that any drastic change in the means of storing and retrieving specialized, systematic scientific and professional knowledge may bring about significant change in the behavior of large segments of the public. The printing press revolutionized knowledge dissemination, storage, and retrieval in the Middle Ages by making duplication, preservation, and accessibility of books much simpler than was the case with hand written manuscripts. The use of electronic computers for information storage and retrieval is again revolutionizing knowledge availability, accessibility, and utilization. With regard to the professions, patterns of knowledge use may well change significantly. The modern knowledge storage and retrieval capacities are likely to support the already existing trend toward greater reliance on more explicit, codified bodies of knowledge. Furthermore, such knowledge is not only accessible to the professional but to others as well. A serious threat to the knowledge monopolies of the professions may arise.

Let us illustrate this point with only one professional domain; namely, that of medicine. For example, automated history taking and computerized physical examinations schedules involving sophisticated technologies are already in use in some major hospital facilities. In some instances, sophisticated, computerized apparatus is currently available capable of directing the treatment regimen of patients without the intervention of professional medical personnel. Modern knowledge retrieval systems offer hope of overcoming the recurrent problem of the absolescence of professional knowledge of medical practitioners. All of these considerations suggest that medical education and practice may be significantly altered.

Some students of professions, especially Wilensky, have observed that narrowing the technical knowledge base of a profession makes it vulnerable to codification into sets of rules and instructions that could be grasped even by those with less training than the full professionals. The routinization and segmentation of professional tasks may be of consequence. At the same time, such developments feed into the growing call for public accountability of professionals. Thus, we now have professional standards review organizations (PSROs) that are monitoring systems to record and evaluate the standards of performance of medical practitioners. Analogously, the 1972 Malpractice Commission suggested that physicians need to be re-examined and relicensed every five years to insure that their professional knowledge remains up to date. Both of these recent innovations are a consequence of the vastly expanded capacity of information storage and retrieval systems to permit the kind of massive data processing and ongoing evaluation of physicians' performance that are increasingly considered necessary in order to maintain standards of competence.

These issues illustrate only one aspect of the potential consequences flowing from the expansion in knowledge storage and retrieval systems. An analysis of the full scope of these developments and their social structural significance remains to be achieved. To give just one indication of the scope of the relevant changes: the number of data terminals in use in the United States grew from 185,000 in 1970 to 800,000 in 1974 (Spence, 1974).

One of the important points about an analysis of these issues must involve the recognition that knowledge is inherently a collective, rather than a private good or property. Knowledge of various kinds, while it is differentially allocated and a scarce resource, is not, like so many other goods, diminished, decreased in value, or consumed in the process of exchange. It is possibly this aspect of knowledge that makes the analysis of its social significance so difficult. We have only begun to touch, in a superficial fashion, some of the intriguing and significant issues that are involved in the storage and retrieval of systematic, organized bodies of knowledge. In concluding this section, we will make a few observations on an extremely elusive matter that somehow lies at the heart of any discussion of knowledge organization, distribution, storage, and retrieval. It is the relation of these social capacities to the cognitive limits in the individual and to collective wisdom, or societal learning.

There is a rather compelling consensus among both social scientists and social critics that there is an outer limit to the amount of stimulation and information that an individual can absorb; postmodern society may rapidly approach this limit or even exceed it (see, for example, George A. Miller, 1956). Indeed, the capacity of individuals to integrate specialized knowledge derived from divergent domains is clearly limited as well. As a consequence, there is no guarantee that increases in specialized knowledge storage and retrieval capacities will lead, necessarily, to wisdom—if wisdom requires the understanding of context and wholes.

Wisdom is holistic, not specialized. The oral traditions, common sense, philosophies, and theology provide one form of such integration. If societal learning means not only the improvement of specialized, technical knowledge, but also the making of wise collective choices, the problem posed is a formidable one. Since the use of available knowledge depends so strongly on interests and purposes, there is no solution to the problem of societal learning that bypasses the political arena. Indeed, the professionalization of the making of choices, as it is proposed in the administrative and policy sciences—that is, removing such matters from political debate and placing them under the jurisdiction of experts—may well be less promising than is often thought. We will return to these considerations after we explore, in the following chapter, the complexities of knowledge use.

8

Knowledge Application

In focusing on the use of knowledge, this chapter examines the manifold interrelations between the knowledge system and the diverse world of practice. Although we believe a reasonable starting point for the analysis of a knowledge system is the function of knowledge production, we realize the relation between knowledge production and knowledge use is complex, especially in differentiated knowledge systems of post-modern societies. The sequence of topics required by linear exposition should not be interpreted as indicating that knowledge use follows necessarily its production or that knowledge is unidirectional in its flow. An argument might well be made to treat the topics in a different order, beginning with the idea of knowledge demand and incentives that are transmitted into the knowledge system through what we call structures of relevance. That procedure, too, would have had its limitations in dealing with a complex, multidimensional and interdependent reality.

In studying explicit knowledge use as the most salient interface between the world of practical activities and the knowledge system, we offer alternatives to prevailing perspectives on the analysis of knowledge systems. Many such analyses are implicitly written from the point of view of knowledge producers; this chapter addresses the matter from the perspective of the knowledge user as well.

We treat the terms *knowledge use*, *application*, or as it is sometimes called *utilization*, as synonymous. With all these terms we refer to actions that explicitly modify practical activity or plans through the introduction of schemes based specifically on organized and structured knowledge taken as factual. There is, then, cognitive primacy in knowledge application. We will attempt to present a clear understanding of the act of knowledge use and its components. Some concrete examples will be helpful in the task. We will then proceed to consider the location of such actions in the knowledge system, the manner in which knowledge resources are brought to bear on problems of action through structures of relevance, and we will conclude the chapter with a consideration of cultural models and institutional designs for knowledge use and knowledge transformations. These matters will lead us into central aspects of contemporary, postmodern society; they are connected with the major dynamics of contemporary cultural and structural change.

THE ACT OF KNOWLEDGE APPLICATION
AND ITS COMPONENTS:
SOME DISTINCTIONS

Knowledge application is a rather special type of symbolically defined or meaningful behavior. Knowledge is brought into play for some instrumental purpose, for the creation of some object or pattern of activity. The focus is on the activity of devising plans for action based on explicit knowledge. This immediately raises the question of how such reflected, purposeful planfulness relates to routinized activity and to the implementation of the plans.

The implementation of plans is an issue of major importance and large scope. Indeed, it is as complex and broad as the issue of deliberate purposes in social change. But these matters differ from the core processes of knowledge use. Certainly, even well-laid plans may run into resistance because they may benefit some people but hurt the interests of others—or are believed to do so. New knowledge needs may arise in the context of implementation. The problem of implementing plans for policies, then, is tied to the question of knowledge utilization but not identical with it. Further, it is an area of such complexity and diffuseness, in many ways not directly tied to the analysis of the knowledge system, that we do not analyze it here.

The distinction between knowledge production and knowledge application is of course of major importance, but it is somewhat less clear than would appear intuitively at first sight. As several examples will show later on, it would be a mistake to believe that knowledge application simply consists in the schematic solution of a problem of action by subsuming it under already established, knowledge-derived rules. There are, to be sure, instances of such off-the-shelf, "canned" applications. In such instances knowledge use has become routinized into a repeatable technique. We find illustrations of this sort of thing in the world of the professions: what is unique and extraordinary to the client who wishes to build a new house probably is routine and recurrent to the architect hired. Indeed, to some extent this is the reason the client turns to the professional in the first place. However, the routinization of technique is not the core of knowledge application in the sense in which we treat the issue. Rather, the core phenomenon is the deliberateness of knowledge use in planning, which always retains an element of the non-routine. Indeed, in the course of practical knowledge use, new knowledge often needs to be gained—sometimes historically new knowledge, and sometimes knowledge new to the participants. In short, we see knowledge application as a creative process having many analogues to knowledge production, the difference being that, in the latter case, the purpose of activity is specifically knowledge focused, whereas, in the former, the focus is on the solution of a problem in practice. The strict differentiation between knowledge production and knowledge utilization is an analytical one and there are major areas of overlap. In the historical transformation of the knowledge system in post-modern society, explicit and deliberate knowledge use becomes an increasingly important phenomenon, and the interdependence between it and knowledge production becomes institutionally closer.

It also must be pointed out that there is a difference between the distribution of knowledge and its application. Processes that distribute knowledge through a social structure by transferring information and the requisite skills are of enormous scope. The entire educational system, particularly contemporary mass education systems, has profound effects on the distribution of knowledge and access to it. Mass media, such as television, newspapers, and the like, play a significant role. Indeed, major efforts are directed toward deliberate knowledge transfers, as in the case of the dissemination of public health or family-planning information to the public at large, or the planful dissemination of technical knowledge to special target groups, such as farmers. The knowledge-oriented, post-modern society is characterized by a proliferation of planned and unplanned channels of knowledge distribution. Indeed, purposeful governmental efforts at knowledge dissemination, while important in their own right, seem to play a subordinate role to the knowledge transfers that occur without plan as a consequence of millions and millions of uncoordinated decisions—a phenomenon usually termed knowledge diffusion.

These processes are of major macrosociological significance in post-modern society since they affect the structure of the knowledge system itself. They also affect the pattern of jurisdictions and authorities within it. Certainly, it is one thing to practice a profession, such as medicine, in a situation where there is an almost complete knowledge monopoly in the hands of the professional, and quite another to practice it in a situation of widely diffused relevant knowledge. It is easier to defend a professional monopoly when the requisite knowledge and skill are role-embedded and thus neither readily transferable nor too easily evaluated by outsiders. This situation is quite different in the explicitly knowledge-based profession relying on codified knowledge and procedure.

In any case, such processes are not identical with those we have in mind when we speak specifically of knowledge application. The prototypical case of knowledge application is one in which a specific problem or issue is construed in such a manner that it is believed to require codified, systematic, and formal knowledge for its solution. This explicit application is different from implicit knowledge use in which role-embedded and experientially acquired skill and judgment predominate over structured and codified knowledge. The difference is well illustrated by the contrast between handicraft production (and problem solving) and the patterns prevailing in high technology. In steel making, for example, much of the reliance on experience and judgment, which could only be acquired by craftsmen over many years, has been replaced by explicit measurement, analysis, and formal scientific prescription. When we speak of explicit knowledge application, we refer to the latter rather than the former.

Yet, such explicit and formal knowledge use always remains embedded in a social fabric in which expertise and judgment continue to play a peculiar role. Indeed, knowledge can intrude in an explicit manner on this fabric in different ways. It may be in the form of measurements, methods, a body of knowledge that permits the mapping of a situation, or broad principles.

In any case, knowledge application is a departure from routine that, however, often aims at routinization. It also is a departure from common sense and yet embedded in it. With these broad ideas in mind, let us cite a few examples.

EXAMPLES OF KNOWLEDGE APPLICATION

We need to turn to examples, concrete instances of knowledge use, in order to push our discussion forward. However, what examples should we choose? The world of practical action is as diverse and colorful as human life itself and knowledge application occurs in all of its domains. Sex and child rearing, education, the solving of social problems, and the life of the professions could supply us with examples. However, we have selected a few instances that stand for many others, in that they illustrate aspects of what happens when deliberate knowledge use occurs. Our examples have another characteristic in common: they all describe innovations. This is so because institutional and technological innovations are the core of the contemporary phenomenon of knowledge use. The systematic planning of organizational designs for innovation—as in the modern research and development complex—is a key feature. And we claim that certain principles distilled from the understanding of knowledge-based innovation will be illuminating even when we turn to instances of knowledge use that deal more with the reduction or absorption of risk and uncertainty than with innovation itself—as in the case of the practicing professions and the world of expertise and consultation.

We begin with an institutional innovation, the planning that contributed to the development of the science-oriented university, which, in turn, became a pivotal basis for the modern research university. The episode is that of the founding of the University of Berlin in 1809–10, which created a prototype novel for its time. The historical background is an immense crisis in the German countries, especially in the Kingdom of Prussia, resulting from defeat by the armies of Napoleon. Existing institutional forms and social arrangements that had been stable, indeed ossified, were now perceived as catastrophically inadequate. The shock of learning that the glorious Prussian army was militarily weak against the French troops went very deep. Beyond that, the idea of progress through rationality, the example of the French Revolution, opened the possibility of searching for alternative institutional forms. The cultural, military, and administrative leadership of Prussia moved increasingly into the hands of people who desired reform.

Universities in Prussia and elsewhere prior to this time had been thought of primarily as professional schools. Indeed, the Prussian universities had been thought of as direct instruments of the state. In the sixteenth century, the Duke of Prussia even threatened recourse to physical punishment in order to quell faculty dissension at the University of Königsberg. Many of the details of university life were regulated by ministerial edicts. The teaching load of the faculty was very high and the general atmosphere certainly not one of creativity and freedom. Even outstanding professors like Immanuel Kant complained about such conditions.

In the years prior to the founding of the University of Berlin, many proposals and memoranda were written on the subject of a new form for the university. Broad public debate on the subject linked specific issues of university reform to conceptions of science and knowledge, on one side, and nationalistic aspirations, on the other. Indeed, the endeavors for university reform were embedded in a broad political and cultural movement. It was from the fervor of the nationalistic movement that much of their intensity derived.

A particularly important document was prepared by Johann Gottlieb Fichte under the title "A Deductive Plan for an Institution of Higher Education to be Established in Berlin Which Would Maintain a Proper Relationship to an Academy of the Sciences."

This memorandum had been commissioned by the government and was written in 1807. It was first published ten years later but played a role in the governmental actions on the new university. Wilhelm von Humboldt, in spite of his brief service in the Prussian cabinet, became the major figure in university reform. However, Fichte was the first rector of the University of Berlin, working to implement a design to which he had contributed.

Fichte's memorandum proceeded from a rather detailed clarification of the conception of the purposes and nature of the proposed institution. Analytically he emphasized the transmission and the storing, but especially the production of new knowledge. The university structure should be designed in such a way that its ability was guaranteed to "be and achieve what no institution other than it can be or achieve, also guaranteeing its own continuity. . . ." From these premises, Fichte deduced the need for a totally inclusive, but also sovereign and autonomous institution. He proceeded to describe in very considerable detail the practical requirements and possibilities, including regulations for budgeting, salary structure, and administrative staff. It is clear the proposal drew on an explicit philosophy of knowledge that was combined with a high level of practical expertise in administrative affairs.

This instance of a social innovation clearly illustrates several points concerning knowledge application. There existed a relevant body of knowledge in the understandings of science and scholarship achieved by philosophers. Moreover, sufficient experience with the administrative and legal flaws of the older system had been critically amassed. One immediate incentive for bringing this new knowledge into play was the acute sense of threat and the need for renovation following military defeat. The context, however, was one of a social reform movement that penetrated the administration of the state and mobilized academic circles. Our brief account does not adequately convey an impression of the turbulence and sense of drama of these events: while the knowledge to be applied was surely rational enough, the forces motivating action derived from the passion of political engagement.

We now turn to a completely different domain of social life and historical period. Here is a brief account of an abortive invention, the story of the copper-cooled automobile engine. We follow the account of the episode given by Alfred P. Sloan, Jr. (1965:71ff). In 1918, Charles F. Kettering started to experiment with an air-cooled automobile engine in his workshop in Dayton. Such an engine in itself was not a novelty. Kettering's idea was to replace the usual cast iron fins for radiating heat with copper ones, since the conductivity of copper was many times that of iron. The execution of this idea, however, required technological knowledge with regard to design and raised basic metallurgical problems. In spite of difficulties posed by the differences in expansion and contraction of the two metals, the prospect of such an engine was attractive. It would simplify automobile construction, increase the reliability of engines by reducing the number of parts, and possibly even improve engine performance. The possibility existed that such an engine might revolutionize the entire industry. This prospect may have been in the background when, in 1919, General Motors purchased the Dayton companies in which Kettering was involved and made him head of General Motors research and development efforts. The frustrations encountered with the copper-cooled engine were influential in the early structuring of that organization.

A difficult conflict soon developed between the research organization and the manufacturing divisions, which paralleled the conflict between the top management of the corporation and the management of the divisions. The central management backed Kettering's commitment to the copper-cooled motor. The divisions that were expected to manufacture it and sell automobiles thus equipped, by contrast, were most skeptical. Central management decided to require that Chevrolet and Oakland proceed with the air-cooled program. Sloan observed that this decision to impose a revolutionary program of work on the divisions ran counter to the management philosophy of General Motors, but he also pointed out:

> In the extenuation of what was done, I should say that this was the first time, to my knowledge, in the history of General Motors that intimate cooperation was called for between the research corporation and the divisions on an important problem, and no established means existed by which this cooperation was to function. Since the initial production as well as the creation of the design was assigned to Mr. Kettering's research group in Dayton and the actual mass production was assigned to the divisions, the responsibilities were blurred. Mr. Zimmerschied wanted to know who was advisor to whom on production: Research to the car division or the car division to Research? Even if there had been no question as to the merits of the new design, this would have represented a problem in management. As it was, there was skepticism at Chevrolet about the new engineering design, and anxiety at the Dayton laboratories that the car divisions would change the design. *   *(Sloan, 1965:76)*

When Kettering's air-cooled car was tested at the Oakland division, the first time it had been subjected to close examination outside Ket-

*© 1964 by Alfred P. Sloan, Jr. Reprinted by permission of Harold Matson Co., Inc.

tering's laboratory, it was found to require a great deal of further development. A deadlock resulted. Even by 1922 General Motors had made no progress towards the new product. Sloan (Ibid.:79–80) wrote about this:

> These events bothered me to the extent that I attempted to raise them to a higher level in my mind with a view to taking them up with the executive committee. I did not feel strongly one way or the other about the technical question of an air-cooled versus a water-cooled engine. That was an engineering matter, for engineers. If I have any opinion today it is that Mr. Kettering may have been right in principle and ahead of his time and that the divisions were right from a development and production standpoint. In other words, in this kind of situation it is possible for the doctors to disagree and still all be right. From a business and management standpoint, however, we were acting at variance with our doctrines. We were, for example, more committed to a particular engineering design than to the broad aims of the enterprise. And we were in the situation of supporting a research position against the judgment of the division men who would in the end have to produce and sell the new car. Meanwhile, obsolescence was overtaking our conventional water-cooled models and there was nothing in the official program to protect their position.

Sloan argued that it had been wrong to entrust not only the principles for the new car to the research group around Kettering, but also its development.

The experience of the manufacturing divisions with the air-cooled engine continued to be poor. In the meantime, Kettering became disenchanted and even threatened to leave the corporation. A compromise solution was worked out whereby the divisions were ultimately relieved of responsibility for implementing a plan they did not believe in. Further work on the copper-cooled engine project was put entirely into Kettering's hands. Finally, new principles for relating an autonomous research effort to the planning and work of the corporation were established. The project itself lost the interest of the corporation and petered out. Sloan emphasized that the experience with the copper-cooled engine demonstrated the difference between advanced product engineering and long-range research, and it painfully demonstrated the need for organized cooperation along clearly structured lines between research and development, advanced product engineering, and manufacturing.

Several points stand out from this illustration. The definition of the problem to be solved arose from a technical prospect, one might even call it a vision, created by a technological entrepreneur, Kettering. However, the actual realization of the plan required more technical knowledge than was available—a sequence of problems arose for which the organization was not structurally prepared. The conflict between the test results obtained in the laboratory and those obtained by the manufacturing division illustrates the significance of what we have come to call reality tests. It appears that quite different criteria were applied at different stages in the process. In the end, management adjudicated the dispute of throwing its weight behind the safely proven conventional design, but at the same time assuring the research arm considerable independence.

Our third example can be described very briefly because it is a well-known and relatively recent event—the invention of the transistor at Bell laboratories. This is an instance of the complex intertwining of knowledge production and knowledge application by purposeful design in a modern industrial research and development effort. There is little doubt that it required considerable organizational resources. Before the availability of transistors, the standard electronic amplifying device was the vacuum tube. While performing similar functions, the transistor has many obvious advantages due to its small size, its ruggedness, and its lack of need for heating. The idea emerged out of a research program dealing with semiconductors. The program, initiated before World War II, was led by William Shockley and S. O. Morgan. Shockley predicted on theoretical grounds that semiconductors could be used to control the flow of electrons in a solid. However, much work was required before a practical device could be developed. The effort was set aside during the war, but immediately resumed by Shockley after its end. A number of basic discoveries of major theoretical significance about the behavior of semiconductors were made at the laboratories. The practicable invention was announced by Bell laboratories in 1941. Since then its further development and refinement have revolutionized many industries. This breakthrough was made possible because of a very systematic understanding of the need for basic research in relation to potential applicability. The experience of this episode in the history of invention is entirely different from that of the copper-cooled engine. It is predicated on the modern industrial research laboratory in which a broad conception of a useful device can be pursued, through organized and cooperative effort, to its conclusion.

Every one of these episodes shows, of course in different ways, the social embeddedness of innovation through knowledge application. Some conception of the problem to be solved, and indeed an anticipation of the solution, arose in early stages of the episodes. The press for knowledge search stemmed from perceptions of opportunities or threats, but in every instance there was some break with routine. Because of this break in routine and the introduction of new perspectives, leadership appears to play a very special role, as in the role of technical entrepreneur, through institutional and organizational factors that mediate the availability of resources. There is little doubt about the significance of chance and uncontrolled, unplanned confluences of ideas and interests in such processes.

Led by the hints gleaned from these examples, we now turn to a more systematic discussion of the act of knowledge application and its components, in which we take our departure from the prototype of knowledge application in innovation. At the same time, we recognize that there are other related types, such as knowledge application in the mode of reflectivity characteristic of much knowledge use in planning, evaluation, and deliberate social change; and knowledge use in the mode of risk absorption that characterizes the pattern of professional expertise.

THE ACT OF APPLYING KNOWLEDGE

Explicit Knowledge and the Break in Routine

An action is a configuration of behaviors or performances that are symbolically defined as a meaningful whole, a comprehensible entity. In this sense an action has boundaries, a beginning and end. It is structured in terms of intentions and plans to carry them out. At the same time, it is a dynamic entity unfolding throughout its course and is rarely entirely predetermined at the onset. As we turn to the specific study of the act of applying knowledge itself, we need to point out that we are dealing with the most specific and, in a sense, elementary level of analyzing the knowledge application problem. We are focusing on the sociologically irreducible elements in the interface between the large structures of the knowledge system and the world of practice. Our subject matter is the activity and accomplishments that unite the recognition of a problem, the search for knowledge, and the means for solution that represent the ultimate conclusion or product. This is the micro arena of knowledge application, situated somewhere in a complex structure that profoundly affects it—for example, through differential knowledge accessibility, command of resources, and the like.

Our examples of institutional or technological innovations show that those directly involved believed themselves to be engaged in an activity that was at least in some sense extraordinary. By definition, innovation is a departure from the routines of everyday life. The break in the routine begins with the perception of the problem, maybe even with the incentive to look for such breaks and possibilities for innovation. In the competitive marketplace and in innovation-oriented institutions, there are massive premiums to be won by discovering departures from routines that allow for novelty. We have seen that there are images of possible accomplishments that can mobilize commitments and the use of resources, as well as initiate knowledge searches.

What is quite remarkable about such acts of knowledge application is the turn to explicit, organized knowledge and the decision to fit it into action contexts, thereby rearranging them. Such a reliance on explicit and codified knowledge is an unlikely event indeed. Most human action does not occur in such a manner. In everyday activity, we rely far more on unreflected action patterns than on deliberate and explicit knowledge. It is the explicitness of codified knowledge that is both its strength and limitation in relation to action. Its strength is that it is learnable and discernible as a cultural object, that it forces reflection and planning. Its weakness is that explicit knowledge needs to be approached on its own terms, so that there is always some gap between it and the problems of action that need to be solved.

Even though it readily can be seen that knowledge use and innovation are disruptive of routine, this may be less apparent in the reflective or in the risk-absorbing pattern. To illustrate this case, we turn to some examples that concern the use of social science in planning and assessing social activity.

Much thought has been given to the problem of how to create new institutions systematically by a group of contemporary scholars who elaborated the concept of *institution building*. Here we have a model for knowledge use that combines innovation with reflectivity while, in a limited sense, involving the expert as an absorber of risk. An interuniversity consortium studied the role of concepts and strategies for building new institutions for social and economic development. Institution building was understood as a particular approach to the problem of bringing about planned change. The collaborators found that American expert advisors to foreign governments had more opportunities to bring a higher level of rational knowledge application to bear on various situations than is usually the case in the United States. Eaton comments on this (1972:11):

> The standards of rationality that are applied in the planning for social change by Americans overseas often seem more professional than many of those advocated for similar domestic projects. In the U.S. domestic scene, technically trained experts usually have staff posts. Few become involved in policy level planning. Experts tend to be suspect by the politicians. The latter feel that their role requires them to adjust long-range planning objectives to the day-to-day short-range perceptions of interest of their constituency. Politicians more often than planners decide when and how to compromise, to adjust what they want to what they think they can get. . . . In the foreign aid field, social scientists are more often trusted to perform both planning roles.

This approach reflects a particular international power system and historical epoch. Much of the institution-building conceptions could be characterized as designs for the administrative absorption of politics. We also have to be cautious about the degree to which institution-building plans actually turned out to be practicable and were implemented. Yet given all of these limitations, this subject represented an interesting conception of knowledge use.

Milton J. Esman (Eaton, 1972:21) described the institution-building perspective as one that "posits purposeful social innovation induced by change-oriented elites who work through formal organizations. Their objective is to build viable and effective organizations which develop support and complementarities in their environment." A comprehensive set of concepts is proposed to describe the "institution-building universe." They include such items as leadership, doctrine, program, resources, internal structure, and various types of linkages with the social environment. The conception excludes autonomous and spontaneous innovations that result in the structuring of new institutions, as well as coercively imposed innovations. The workability of this particular conception of knowledge application is entirely dependent on "some measure of official sponsorship or indulgence" for the change agent. Given this vantage point the effort is guided by a theory of institutionalization.

> About *change processes* the model implies induced rather than spontaneous initiation and guided rather than autonomous diffusion. Unanticipated events, some favorable and some unfavorable to change agent goals, will inevitably crop us and these must at times be dealt with expediently; but in this rationalistic guidance model, the main reliance is on planned and managed change. The three main change processes are technological, cultural, and political. *Cultural* or normative methods rely on methods to change individual or group values, attitudes, or role perceptions using ideological, indoctrinative, emotional, symbolic, group dynamic and other sub-cognitive methods. *Technological* methods rely on cognitive information or on new practices or services to induce fresh action patterns and intellectual commitments to changed roles and activities. *Political* methods rely on the redistribution of power, redefinition of rewards, manipulation of resources, or the use of influence and arguing to produce behavioral change. In any major IB effort, all three methods must be used in a variety of sequences and combinations.
>
> *(Eaton, 1972:28)*

This obviously is not a comprehensive model of social change as such but, rather, a design for an instrumental approach given a particular set of modernizing objectives. The conception is highly rationalistic and reflective. The emphasis on reflectivity means that knowledge application is not merely conceived of as the use of established principle for the purpose of designing a new plan, but knowledge application is also thought of as continuous monitoring and assessment for the purpose of making an otherwise opaque situation transparent.

This becomes particularly clear in the statement by Jiri Nehnevajsa (1972) on "Methodological Issues in Institution Building." He views institution-building research as an integral component of the process itself. It inevitably involves anticipations of the future. There should be attention both to a design and an evaluation cycle. The evaluation cycle requires the explicit identification of goals and their analysis leads to the study of their eventual realization. The design cycle runs from conception through various assessments to steps for redesign. This kind of reflectivity obviously requires continuous monitoring. Here we have a conception of the policy-making process in the specific area of social or institutional innovations that suggests continuous social measurement for the purpose of assessment and redesign in the interest of improvement.

Even though our example is a specific one, the broad ideas underlying it have found wide utilization. Institutionalized social measurement for the purpose of monitoring major social functions, such as in the construction of social indicators, is a well-established aspect of knowledge use in contemporary life. Indeed, measurements and the properties of indicators often come to be compelling sources of definition for new problems requiring action. For example, the methods of business accounting monitor the economic health of a firm; similarly measurements of economic performance, unemployment, and polls on the acceptance of government policies are presented as measuring a society's well-being. The dimensions of such measurements, in turn, become codeterminants of the detection and construction of problems. This is an aspect that we will return to later when we deal with the process of problem construction. It is one indication that reflectivity in knowledge use is both enabling and constraining.

A somewhat different perspective on the relation between knowledge use and routine activity is found in the conception of planned change through organizational consultation. We are referring here to an intellectual tradition stimulated by Kurt Lewin's conception of action research. Much attention is given to the role of groups and their internal dynamics in bringing about both personal and organizational change. In this context, the role of the social scientist as a source of social knowledge is less instrumentally conceived than in the institution-building approach; more attention is given to the application of social knowledge in stimulating self-reflection and spontaneous improvement.

Chin (1966) suggests a typology of design and utility factors according to the use of social research in the context of change efforts. He distinguishes six types: evaluative research, which assesses the effectiveness of a program; predictive research, which attempts to provide information relevant to the making of decisions; technique research, in which a particular method is tested so as to assure its utility; action research, which is intended to affect conduct by providing information not normally available to relevant actors; practitioner theory research, which is oriented toward the building of scientifically valid principles for professional practice; and, science building research, which aims at fundamental scientific theory. These distinctions are based on different relations between the knowledge provided by social research and the requirements of practical action. Knowledge application in this conception is thought of as a mixture between providing reflective information for the assessment of ongoing activities, fundamental principles, and the construction of bodies of knowledge specifically designed for the purpose of repeated applications in professional conduct.

In an earlier context, we referred to Paul Lazarsfeld's (1975) conception of the "utilization cycle." This conception illustrates various breaks in routine involved in research utilization. Six stages are identified in the utilization of social research. They begin with the formulation of the problem (stage 1); the setting up of staff and forming of an organization for knowledge collection (stage 2); the search for knowledge (stage 3); the forming of recommendations (stage 4); implementation (stage 5); and assessment (stage 6), after which the entire cycle may start again. This bears a certain similarity to the conception of "monitoring" formulated by Nehnevajsa in the context of institution-building research. The Lazarsfeld formulation suggests that there are two important gaps in this process of knowledge utilization: the gap between the problem as stated by men of action and the problem as formulated for research, and the gap between the knowledge assembled and the formulation of practicable recommendations. Both of these gaps involve transformations, that is, changes in perspective. Very little of a systematic nature is said about how to eliminate these gaps. However, Lazarsfeld and Reitz do offer suggestions as to what type of considerations should inform the judgment of professionals confronting gap problems of this kind, so that their dilemmas and difficulties are reduced.

Another gap in the application of social knowledge is pointed out by Argyris who emphasizes the importance of the gap between emotional and cognitive realities (1972):

> Recently several case studies have been published of top management groups that attempted to introduce structural changes before they altered the degree of openness, risk-taking, and experimenting in their interpersonal and group dynamics. The data suggest that the structural changes were neither fully developed nor effectively introduced. One of the biggest bottlenecks was the interpersonal and group dynamics of the top groups. Their dynamics led to members having their ideas about structural changes undercut, . . . therefore leading to minimal sense of clarity of agreement with understandably less internal commitment to any given structural change. . . .

Argyris did not intend to substitute a focus on emotionality and group dynamics for rational analysis of environmental pressures and technical considerations. However, he did argue that the effective adoption of available social knowledge and plans based on it require recognition of emotional factors and a constant readiness to deal with these noncognitive factors in human interaction.

These approaches to the utilization of social knowledge complement what we have learned from the study of institutional and technological innovation. But another element must be considered: namely, the absorption or transfer of risk. This aspect is best illustrated in the world of the professions. Professional practice involves the continuous and systematic application of knowledge and expertise. Here knowledge application, which we have discussed from the point of view of its extraordinariness, its break in routine, has become routinized for the professional, but is extraordinary and problematic for the client. Eliot Freidson (1970:203) has described this in subtle detail as the "social construction of illness." In the course of medical diagnosis and treatment, a sick person's status is not only redefined to that of patient, but one's entire life situation may be reconstructed in terms of medical categories and concepts. The central aspect of this process is the authoritative absorption of uncertainty by the physician, based on the presumption of the physician's application of valid knowledge. In this context, the patient must assume the status of an object in a situation defined by the physician. The patient's trust in the physician is contingent on a belief in competence; through such trust, the extraordinary aspects of the patient's experience become, as much as possible, managed routinely by the medical profession's norms. In this particular type of knowledge application process, the key element is the unequal reciprocity of perspectives; what is extraordinary for the patient may well be routine for the physician. At the same time, this process involves a transfer of uncertainty that to some degree appears inherent in any knowledge-based practice involving clients or consultation.

Renée Fox (1957) has called attention to provisions made in professional training for dealing with uncertainty. Uncertainties arise because of the limitations of professional knowledge, because of the limitations of a particular practitioner's mastery of this knowledge, and, in a particularly difficult manner, because of lack of fit between these two sources of uncertainty. She shows that in the course of medical training students are increasingly taught to recognize these kinds of uncertainties and learn how to come to terms with them. One difficulty is the maintenance of a reassuring stance towards patients while dealing with medical uncertainty. In domains in which knowledge application is not necessarily transparent to the user—and medicine is certainly one such domain—trust in expert authority is particularly central and vulnerable. In addition to the aspect of instrumental problem solving through knowledge use, the element of uncertainty transfers enters dramatically in these domains.

A similar point can be made by referring to the practice of law. Here, too, troubles or disputes that may well be unique and extraordinarily disruptive for the individual involved are best dealt with by transferring responsibility to the professional practitioner. A lawyer is presumed to know not only the substance of the law and its procedures, but also to be an expert in assessing what can concretely be done for a client. Legal training, like the socialization process in other practicing professions, involves preparation for such uncertainty transfers. One particularly interesting indication of this can be found in the often noted conflict between lawyers and social scientists on presidential commissions inquiring into major social problems. In some instances, it appeared that the social scientist wished to conduct research that might have the effect of increasing uncertainty for policy makers, whereas lawyers were inclined to reduce the problem to manageable proportions and relate their knowledge search to actions known to be possible with the existing governmental machinery. Lawyers also have been noted to be quite effective in responding quickly to a call for service, being organizationally and professionally prepared to accept what for others are extraordinary assignments. The cultural and institutional model of the practicing profession, then, provides for very specific and peculiar modes of knowledge application. Bodies of knowledge are related to, but not identical with, principles of professional practice that are embedded in contexts of authorization and legitimation that enable the professional to exercise situation control, in the sense of limiting the scope of the situation to be dealt with, in exchange for which the professional assumes responsibility for reducing client uncertainties. The professional, both in medicine and law, is expected to be a link not only between knowledge and advice, but also between the formulation of concrete plans and their execution.

Common Sense and Role-Embedded Knowledge

We have spoken of common sense as a cultural system that is the point of departure for all specialized modes of knowing. These specialized modes of knowing may, however, depart greatly from it. Yet the cultural system of common sense remains an embedding context that patterns specialized knowledge use. Common sense represents a socially anchored structure of resources for the interpretation of situations that is broadly shared in a society. Its core is believed to represent reality itself, as well as guides or principles for action. Such a system defines pathways for thought and action. Our meaning of common sense is very close to Alfred Schutz's conception of the "paramount reality" that the wide-awake person encounters in everyday life. Indeed, Schutz has given a rather sensible description of the structure of this paramount reality, with its compelling hardness and its temporal and spatial dimensions.

Common sense guides and constrains thought and action in everyday life. It provides for routines. At the same time, there are numerous highly differentiated patterns and specialized routines. Max Weber also emphasized the sociological notion of everyday life, which he contrasted with the extraordinary state of "charisma" that is always disruptive of routines:

> The term "charisma" will be applied to a certain quality of an individual personality by virtue of which he is considered extraordinary and treated as endowed with supernatural, superhuman, or at least specifically exceptional powers or qualities. These are such as are not accessible to the ordinary person, but are regarded as of divine origin or as exemplary, and on the basis of them the individual concerned is treated as a "leader."
>
> *(Weber, 1968: vol. 1, 241)*

Weber conceived of the nature of everyday life as molded by economic terms. There are economic constraints on the lives of family and household units and these compel work and exchange to occur in recurrent, routine patterns. His conception of everyday life as routine social activity emphasized people going about their daily affairs of making a living in the context of a web of social relationships regulated partly by interest constellations, partly by norms and values, and partly by established authority. Charisma appears where the institutional structure of this web of relations is disrupted, as occurred in the great pressure zones of history during the rise and decline of empires through political, technological, or religious change. Under these circumstances the nonroutine intrudes.

The excitement and enchantment of the charismatic moment inevitably cannot last. The thought patterns of common sense and the routines of daily life reassert themselves and compel the "routinization of charisma."

Weber's conception of everyday life is clearly juxtaposed to the notion of the extraordinary. The former emphasizes economic exigencies, routines, action in the context of the familiar and ordinary. The latter tends in the direction of the exceptional, the supernatural, that which transcends the world of paramount reality, the sacred and holy.

Weber's ambivalent attitude towards rationality and rationalization is expressed in his fear that it would lead to an "iron cage" of bureaucratic regulation, the ultimate victory of the routines of everyday life through the disenchantment of the world. In this vision, every-thing would become grey and ordinary. This imagery does not mesh with what we find in the process of knowledge application. It is quite mistaken to conceive of knowledge application as if it were a straight-forward matter of calculation and rationalization, itself routine. On the contrary, it involves the challenge of dealing with the extra-ordinary. In every instance in which the application of explicit and organized knowledge is sought, a break in routine must have oc-curred, at least to the extent that the existence of a problem requir-ing information and judgment and the explicit formation of plans is recognized. Here the resources of routine activity, relying on role-embedded knowledge and commonsensical guides for action, which require little if any explicit reflection, have become insufficient.

The extraordinary opportunity, challenge, disaster, or simply the desire for remedial improvement of something, requires a cogni-tive distance from patterns of routine activity. From within them, the problem is not recognizable in the first place, except possibly as a puzzling obstacle. If the cultural system of common sense supports what is to be taken for granted, then explicit knowledge application is an important departure from it.

Certainly the element of extraordinariness in the action of knowledge application is different from the emotionalized intensity of commitment in Weber's charismatic movements. And yet there remains in it a sense of adventure, the tension of grappling with the exceptional that needs to be noted. Indeed, we will find that the directions of knowledge application are much influenced by great mobilizations and movements that help shape the social agenda of what is perceived as a problem and where the breaks in routine and the confrontation with the extraordinary occur.

Departures from the ordinary, both in the direction of tran-scendence and in the direction of innovative use of knowledge, occur repeatedly, and the cultural system of common sense provides link-ages and pathways to these domains. Common sense not merely defines the paramount reality of objects as things, but also sets standards of prudence and good sense, typically embedded in the knowledge contained in ordinary role definitions.

Of course not all knowledge use is of the explicit and codi-fied variety. Much knowledge use relies on implicit, role-embedded knowledge and skill that has to be learned through apprenticeship or incumbency. Such role-embedded knowledge, expertise in the broadest sense, is the pattern in the typical occupation and especially in the professions. It provides resources for judgment through which various gaps in explicit knowledge use are filled and knowledge transformations may be accomplished.

The Break in Routine and the Construction of the Problem

Problems requiring knowledge application need to be recognized and diagnosed in the first place. This is a very difficult process that requires a break in routines. An interesting discussion of the issue occurs in Lazarsfeld and Reitz's book on applied sociology (1975:48ff):

> "Problems" are not as clear-cut as first it may appear. People have to become aware of them in some way, and before a problem can be translated into a research design, an extended dialogue between the sociologist and the man of action will be necessary.

They then turn to the question of how organizations become aware of problems facing them (Ibid:48ff).

> One possibility is a catastrophic event, such as a breakdown of operations or an exposé by a newspaper. It is puzzling that often such events seem to occur before a problem is even recognized from within, but a more detailed analysis will show how difficult it is for top management of an organization to be familiar with every operative detail.*

They describe three orderly ways of discovering problems other than the disastrous event. In every one of these a high level of reflectivity seems to be required. They describe formal monitoring of activities, systematic reporting as by correspondence, and attempts at planned forecasting of future difficulties as ways to detect problems.

Interestingly, Lazarsfeld and Reitz introduce the concept of "half knowledge" (Ibid.:48–49):

*Reprinted by permission of the publisher.

Managers are sufficiently aware of uneasy aspects of their operations, so that they can swiftly mobilize for action. If everything goes well or if danger signals are sent, this half knowledge has great economic advantages. But if it is used as a screen to block out potentially unpleasant situations, it can become dysfunctional. For example, executives of American television networks were greatly disturbed by the quiz program scandals, although they must have had half knowledge of the deceptions. The U.S. space administration was certainly embarrassed when it became known that long before the Apollo tragedy, the chief contractors' work had been criticized, but the project leadership did not take full cognizance of it. Doubtless the auto industry knew of cars "unsafe" at any speed; by blocking out such knowledge it saved money in the short run, though it risked much more stringent control of the industry in the long run.

Half knowledge is a form of routinization that puts into the dim distance what appears unnecessary or even threatening to know. All the forms of problem detection described by Lazarsfeld and Reitz require a break from routine, in the sense of not permitting routinization to enmesh critical reassessment. The virtue of monitoring is, of course, the development of social measurements that invite scrutiny from different perspectives.

A problem emerges where there are not only breaks in routine, but also deficits in the common sense resources for repairing the break at once. This type of situation invites a search for a new configuration of linkages between cognitive elements that had not been seen together before. It is here that knowledge transformations occur through the interplay of different perspectives. This departure of the frame of reference from the routine is one reason why knowledge transformations and gaps between perspectives are such prominent aspects of the knowledge application process.

The Construction of the Problem as Play

It may seem strange to speak of knowledge application as play when it typically involves heavy and sustained labor. Yet playfulness does indeed appear. There must have been pervasive elements of playfulness in the carnival atmosphere surrounding the first experimentation with balloon flights in eighteenth century France. When the brothers Montgolfier discovered that the principles involved in the movement of clouds and in the rising of smoke—the difference in weight between warm and cold air—could be used to lift balloons, they did not appear to have pursued a matter of great immediate practical import. Their first demonstrations were spectacular performances. Only gradually did the idea of balloon flight become transformed into utilitarian devices.

This element of playful exploration for its own sake frequently appears in the phase of problem recognition. The notion of a modern manager "playing around" with different solutions to a problem in attempts to suspend, at least for the moment, the press of serious consequences recognizes this element. The appearance of gaming exercises, in which an atmosphere of the "as if" is created, capitalizes on the point. Playfulness reduces the commitment to a singular point of view and allows the exploration of the unusual. The definition and redefinition of a problem in this playful mode may, indeed, be an element in the search for solution. We suggest that such a phase of playfulness is an important ingredient of creative knowledge application.

The Construction of the Problem as Exchange

Elements with critically serious consequences, however, assert themselves as well. There are premiums to be earned and costs to be minimized in devising a construction of the problem to be faced. Let us illustrate this point first with regard to the controversy involved in attempts to define the nature of a given social problem. Certain issues are presumed to demand policy attention. The manner in which they are defined is significant in offering initial advantages or disadvantages to contending parties. Interest groups will enter the arena in an attempt to affix an interpretation advantageous to their points of view. For example, in recent years, the issue of energy use has been treated from often diametrically opposed points of view. If one proceeds from the assumption that the problem is a deficit in energy supply, then efforts to build effective nuclear power plants are necessary and even heroic. Certainly, the scientists and engineers involved in the movement to build a nuclear-generating capacity thought of the problem and of themselves in this way. If the argument is made that the problem is wastefulness and exploitativeness in energy use, the heroes of the former scenario become villains.

Often the construction and description of a social problem already contain notions as to what might be done about it. There is an incentive for interested parties to attempt to articulate such problems in a manner that arouses the support of the potential allies while minimizing political costs. Clearly, the politics of problem definition is a major and formidable element in the action of knowledge application.

A relatively recent illustration of this point is the successful effort in the United States to redefine arthritis from a purely medical problem into a social problem. Surely, suffering from arthritis precedes recorded history. However, it is difficult to believe that such suffering was defined as a social problem—rather than a curse or a disease—until quite recently. The successful campaign to enlist the resources of the government through the creation of a large-scale arthritis-related research program was certainly political. Redefining the problem involved demonstrating that the disease not only affected large numbers of Americans but also caused very substantial economic losses, and that its unequal distribution in the population raised moral problems concerning distributive justice in the access to healing knowledge.

Considerable rewards may be in store for those who win contested problem definitions; they succeed in the construction of a problem they can ameliorate if given appropriate resources.

The construction of social problems is closely related to the construction of grievances and the detection of dissatisfactions. Such circumstances are not simply and objectively there, in the given reality, but constructed through the aid of cultural models. It is fairly clear that an historical progression has occurred that increasingly removes dissatisfactions and grievances from the domain of moral sanction and relates them to the domain of planful, instrumental, and knowledge-based action. This can be seen, for example, in the changing attitudes toward unemployment.

The politics of problem definition, then, involve establishing the legitimacy of the problem by gaining the widest possible or the most effective support for one's proposed conception. The struggle for establishing such definitions is conceived of as an arena within which demands for interpretation generate a supply of interpretive schemes. Gains at this stage for one of the participating actors may have a significant influence on the subsequent mode of problem definition and solution attempts. The successful definition of a number of American social problems as medical issues may serve as an illustration of this point. In contrast to the Soviet Union or Japan, the United States has increasingly defined both alcoholism and drug addiction as strictly medical problems. This is so in spite of the fact that medical science offers little reliable and useful knowledge for the cure of these diseases—if they are in fact diseases at all. In consequence, the mobilization of knowledge resources and search activities preferentially involve the health-related professions. The fact that such a circumstance is not without advantage in the system of interest groups is obvious.

Considerations of political exchange and advantage may influence attempts at problem definition in an even more basic and narrow sense. In the immediate context of the emerging problem, a premium may be attached to detecting problems that superiors or competitors have overlooked or have improperly defined. This motivation quite often is involved in "young Turk" movements. At the same time, calculated compliance with authority or outright fear may play a role in limiting problem definitions.

Economic exchange and market opportunities offer probably the most important incentives for knowledge application in market economies. Nelson, Peck, and Kalachek (1967:28ff) discuss this matter with particular emphasis on technology. They argue that there are two broad factors at work: differences and changes in the opportunities for reward from different kinds of technical advance, and change in the possible resources with regard to materials and knowledge. Principally, they see economic incentives for knowledge application to be similar to "more pedestrian products and services." On the side of demands or incentives for knowledge application, they give the following example (Nelson, Peck, and Kalachek, 1967:29):

The effects of changing patterns of demand on invention is illustrated by the sequence of major textile inventions in England during the late 18th and early 19th centuries. During the early 18th century the invention of the flying shuttle and the silk throwing machine greatly increased the amount of cloth and knit goods that could be produced per worker. The result was a fall in the price of cloth, more cloth output, and more demand for yarn to make cloth, thus raising economic returns to technical advances in the spinning processes. The profit prospects for successful invention were further enhanced by labor shortages and rising wages since the supply of spinners increased slowly. The work which led to the water frame, the spinning jenny and, later on, the spinning mule (which combined aspects of the water frame and the jenny) was directly stimulated by this increase in the demand for yarn and in wages of spinners. This spurt of induced invention in spinning eventually overshot, and shortages of weavers began to materialize. These new shortages were met in part by a shift of labor into weaving, and in part by the shift of inventors from spinning to weaving resulting in the development of the power loom.

They point out that in this and other examples the demand for innovation is related to two factors. One involves the increases in the demand for the product. The other concerns some shortage in a factor of production that makes for incentives to mitigate the effect of that shortage.

The idea of economic incentives and exclusiveness of use rights is embodied in the patent system. It is widely believed that the patent system encourages innovative knowledge application by granting a monopoly to the patent holder on the exploitation of an invention. Fritz Machlup (1962:176) expresses skepticism on this point. He feels that neither the claim that the patent system promotes innovation nor its denial can be proven:

> Advocates of patent protection have for centuries propounded the faith in this institution, and their statements admit of not an iota of doubt. They may well have the truth—but faith alone, not evidence, supports it.

At the same time Machlup documents the enormous economic incentives linked to investment in systematic research and development.

The Construction of the Problem as
Cognitive Work

Whether the problem is constructed as an issue, a social problem, or a technological task, if it is believed to involve the explicit use of knowledge, it necessarily requires that specialized reference frames and their cognitive skills and knowledge resources be brought into play. In a technical sense, the problem constructed in the arena of politics, economic exchange, or elsewhere needs to be redefined to become amenable to technical treatment. At this point, certain transformations or gaps will occur. There are, of course, public policy problems of such broad scope that they cannot be translated into issues for research without loss. Further, some issues may be phrased in such a manner that they are not sensitive to technical knowledge application.

The specialized construction put on an issue may depend on the prestige of professions or disciplines. The tendency to see certain issues of policy as basically legal or economic in nature, or to classify issues as medical ones, illustrates this point. Indeed, the classification of an issue so that it enters the cognitive jurisdiction of a particular epistemic community is one of the crucial aspects of knowledge application. Entirely different groups and cultural domains are brought into play if, for example, alcoholism is defined as a moral or legal problem rather than a medical one.

In our treatment of reference frames, we discussed some of the differences in cognitive style found in different epistemic communities. Differing reference frames will typically require knowledge transformations and rearrangements in the process of problem construction. We will return to this matter when we discuss the multiplicity of frames of reference involved in the various gaps and in the stages of knowledge use.

Knowledge Use as Objectification

In dealing with the break from routine, we mentioned both changes in frames of reference and changes in the configuration to be dealt with. The conscious and deliberate reflection on a situation, which otherwise would not become the focus of scrutiny, is a significant general aspect of the knowledge application process. It generates not merely the need for accessible bodies of knowledge on which one can draw but, quite typically, the need for systematic, methodical measurement and gathering of facts in situations that otherwise would remain implicit contexts of action, rather than foci of analysis. This implies an extension of the scope for potential instrumental action. More importantly, it is a process of objectification, a process through which circumstances to be dealt with are increasingly removed from the domain of belief and subjectivity and located in the domain of constructed objects and their systematic observation. This process historically has received very ambivalent evaluations. There is the romantic attitude bemoaning the loss of naïve directness, a Rousseauian yearning for return to the purity of the unreflected life. On the other hand, we find the rationalistic cult of technocracy, celebrating the positive age in which human beings finally achieve reflective mastery over the environment.

However one evaluates the phenomenon, it is clear that the reflexivity inherent in the act of knowledge application and the impulse to measure and observe systematically have profound socio-psychological and sociostructural consequences. Indeed, a change from a conception of society as a given moral framework to a conception of collective reflectivity, policy formation, and even collective intelligence mirrors the phenomenon on the societal level.

The Context of Social Structure

Any given act of knowledge application is shaped both in the enabling and limiting sense by location in the social structure. The concentration of channels of communication and resource around certain organizations clearly makes a major difference. The relationship between the social location of those who define a problem and those who attempt to supply knowledge to solve it can affect the process.

Planful designs of organizational arrangements for knowledge use have increased in significance in recent decades; even so, the unplanned, even accidental, confluence of ideas, people, and resources remains a major factor. Social structure can be thought of as a determinant of such probabilities.

Finally, a particular knowledge-based solution to a problem is likely to affect different groups or segments of a social structure differentially. It is a rare circumstance that everyone benefits equally. The multiplicity of interests typically involved means that there are gainers and losers. This is complicated by the different problems involved in implementation and different conditions of routinization. After all, implementing a plan means incorporating it into a pattern of ongoing activity, routinizing it so that it is no longer an object of reflection but a taken-for-granted activity. This circumstance highlights the point we made earlier; namely, that knowledge application is in a peculiar manner deroutinizing and reflective.

In most cases of knowledge application, not all consequences and their reverberations through the social structure are taken into account. There is a natural tendency for actors to minimize costs to themselves. In implementing knowledge-based plans, this can be done by leaving those consequences that affect others unexplored, so that these others bear the cost. Externalizing costs so that they are born by groups not involved in the plan itself or by the general public has become increasingly troublesome in post-modern society. Indeed, the environmentalist movement, with its pressure for environmental and social impact analyses of new developments, can be interpreted as an endeavor to limit such unequal externalization of costs. This results in a press for increasing inclusiveness and reflection in the policy process.

Gaps and Discontinuities Between
Reference Frames and Reality Tests

In a complex issue, knowledge application involves a multiplicity of specialized reference frames varying in degree of cooperation and conflict. Innovative solutions to problems seem to be easier for those that have a perspective at least slightly different from the one customarily involved in the matter. This often seems to be a function of marginal status. We turn to invention as an example of knowledge application and draw on the report prepared by the Batell-Columbus Laboratory for the National Science Foundation. (Batell-Columbus Laboratory, 1973) The case histories of major inventions prepared in this study lead the authors to argue that a "technical entrepreneur" is a most important driving force in the knowledge utilization process. One aspect of his/her role appears to be the management of multiple frames of reference, the maintenance of motivations, and the ability to put together a functioning task group of diverse collaborators.

Discontinuities and differences between reference frames become particularly important in the knowledge-use process. For example, the difference in perspective between policy makers and scientists has been noted often and deplored. This discontinuity results not only from different frames of reference, but also from the gap between action and knowledge in the formation of the problem. Clearly, the first recognition of a problem as a break in routine, and the pressures for its various constructions deriving from playful exploration, or massive political, economic, or other considerations, will not make its treatment obvious in terms of already existing cognitive reference frames. On the contrary, there may well be jurisdictional problems as to which epistemic community and specialized perspective ought to be involved. This transforms the matter into the language of boundaries and perogatives of specialized communities. Lazarsfeld and Reitz described sociologists addressing practical issues and dwelt at length on the difficulties involved in changing an action issue into a research problem. The same gap between active frames of reference and relevant cognitive reference frames is at work whatever the body of knowledge involved. Further, there is the gap between relevant knowledge and the making of plans or recommendations for plans. It is impossible to eliminate an element of judgment and valuation when knowledge is transformed from the cognitive frame back into the action frame. In addition, rarely are all potential difficulties foreseen.

These gaps and transformations between stages of knowledge use and frames of reference pose a particularly significant issue in post-modern society. The proliferation of symbolically distinct bodies of knowledge and domains of expertise seems to require particular attention to the manner in which alternative perspectives and divergent skills can be effectively interrelated. Considerations such as the significance of a theory of knowledge for practical planning and management purposes arise. Certainly the skill to gain access to other frames of reference and to construct adequate maps of their interrelation is a vitally significant one under the circumstances.

The Assessment of Trustworthiness

The search for trustworthy knowledge and experts representing it is, for someone wishing to solve a practical problem, a matter of grave significance. Such a search could not possibly occur in terms of purely cognitive criteria. The searcher would already have to possess an independent evaluation of the knowledge he/she hopes to gain. Moreover, those in need of an urgent solution to a problem cannot be expected to check the empirical and rational validity of all potentially relevant knowledge.

Yet judgments of trustworthiness are made, often on the basis of the authority and prestige of the knowledge source or on the basis of experience. This should not be misunderstood as an irrational procedure. The social distribution of trust in knowledge sources can be taken as an imperfect but useful indicator of how to proceed with the reduction of uncertainty. While few systematic reality tests can be applied in the process of knowledge search itself, the authority of a particular body of professional knowledge rests in its long-range vindication by proving its mettle in practice. Nevertheless, self-validating knowledge use systems often develop that are effectively shielded from pragmatic or empirical disconformation. This may involve the construction of a self-validating subculture. In the case of the psychoanalytic subculture, patients as well as therapists share a similar language and preference for modes of explanation. It is not unreasonable to argue that, under these circumstances, a self-validating loop is built into the structure. Richard L. Henshel (1975:92–106) has recently argued that "a discipline with sufficient prestige can eventually shape the institutional forms of its subject matter, and so far pervade the thinking on the subject that many of its postulates appear obvious a priori." Henshel further discussed certain multiplier and oscillation affects resulting from disciplinary prestige. For example, a particular prediction may be taken as a warning so that it, in effect, brings about its opposite. Or a prediction may become a self-fulfilling prophecy.

Such consequences of social structural arrangements, such as the prestige of a knowledge source, for the assessment of credibility cannot be ignored. This question of trust in knowledge and the credibility of experts cannot, however, be treated merely as the result of rational endeavors and systematic knowledge searches. More profound commitments, hopes, and fears play their roles.

Motives and Identities

Our reflection on the act of applying knowledge has shown it to be anything but mechanical. Indeed, while we have used examples from the domain of technology, we have avoided simplistic engineering notions of knowledge application, as if it were a simple matter of routine rationalizations. The search for answers to problems posed by breaks in routine, the confrontation with the extraordinary that is characteristic of the technical entrepreneur, the fervor involved in the construction of social problems, and the close relation of the knowledge application process to fundamental commitments and values have become apparent. Indeed, when we speak of the interplay between frames of reference and epistemic communities, we use the language of the sociology of knowledge to deal with phenomena of considerable intensity. Such perspectives are backed by conceptions of identities, motives, and commitments. Later in this chapter we will take up the relation between knowledge application and movement commitments. Here we need only point out that the act of knowledge application affects personal and collective identities, as well as the forces that maintain and alter them. This is true not only through the reverberating external consequences of technological or social change, but also through the requirements for measurement, reflection, and objectification that characterize knowledge application itself. This aspect of change is distinct from the technical alteration of external circumstances. Post-modern knowledge application is a phenomenon not at all external to the fabric of society; it alters not only the content, but also the structure of consciousness and thereby the sense of identity.

The Knowledge System and
Other Contexts

Larger contexts than the knowledge system shape the possibilities of knowledge use. For example, the pattern of jurisdictions and contests over problem assignments within the knowledge system at large becomes a significant constraint for the way in which situations of knowledge demand may be linked to sources of knowledge, either in organized bodies of knowledge through information retrieval or in knowledge production centers or knowledge-related services. The debates about multidisciplinarity may serve as an illustration. There is little doubt that the disciplinary model of knowledge production and storage is still the dominant one and that this pattern of organization restricts knowledge utilization possibilities. Decades of organized attempts to institute multidisciplinarity show the difficulty. In the American university, disciplinarity is ingrained in the traditions and structures of departmental organizations; the establishment of multidisciplinary research centers outside the departmentalized domain of faculties of arts and sciences, for example, has been only partly successful. Often such centers simply result in an unintegrated and uncoordinated array of several disciplinary efforts. Status and authority differences as well as differences in frames of reference remain powerful limiting conditions. Efforts at multidisciplinarity have been more successful when they led to the establishment of new disciplines, as in the examples of biochemistry, information science, astrophysics, and others. The core structure of the system shapes the organized distribution of knowledge resources and skills in such a manner that application capabilities remain restricted.

The contemporary American pattern contains organized and sometimes very elaborate linkage structures for the delivery of usable knowledge to points of implementation. Quite different styles of institutionalization have evolved. For example, in agricultural knowledge production and diffusion, there is a fairly effective link between practice and knowledge production. A major university-based effort in the land grant colleges interacts with the farming industry through an institutionalized knowledge distribution network and patterns of occupational behavior. A quite different system has evolved in the aerospace domain where highly sophisticated scientific and technological projects are often coordinated across several firms and research institutes. All of these domains possess to some extent planned and purposeful attempts to manage knowledge production and use, but the degree to which such efforts encompass the entire sector varies greatly. Nowhere has deliberate policy information and knowledge management replaced the effect of unplanned knowledge activity.

We also need to remind ourselves that the priorities and substantive interests in different domains of the knowledge system do not necessarily fit together. In our discussion of knowledge production and growth, we showed how priorities emerged that are inherent to that component of the system. The highly diverse domains of knowledge use in practice generate knowledge demands that may clash with knowledge production priorities. For example, the recent, broad debate about redirecting research efforts so that they fit practical needs, which led to the establishment of a division in the National Science Foundation for research applied to national needs (RANN), illustrates the difficulty. Special criteria and incentives were set up to induce the scientific community to address their work to practical knowledge priorities, rather than those arising from the internal processes of the disciplines themselves. The effort was received with some skepticism.

These episodes serve to remind us that a knowledge system can be thought of as quite differentially responsive to internal as against external incentives. This kind of responsiveness is influenced by the differential conductivity of the system. For example, the segmentation into disciplinary structures within knowledge production and the segmentation between knowledge and other domains often leads to barriers in the flow of information through the system.

This reminder emphasizes that the larger contexts of every act of knowledge application limit the scope of available resources. Indeed, from the point of view of the user, the knowledge system offers facilitating pathways as well as constraints on knowledge search. We now turn to this matter.

Structures of Relevance:
Pathways to Knowledge Resources

In the process of defining a problem and searching for relevant information, each potential knowledge user attempts to find resources that have some constructive bearing on the issue at hand (Holzner, 1973b).

We know something about the natural history of the emergence of issues, of action problems, or grievances. We know, however, not enough about the socio-cultural structure of the processes through which a particular question or issue emerges as the pivotal focus of uncertainty which generates a demand for knowledge application.

We will call these arrangements through which such foci emerge "structures of relevance." By this term we refer to the arrangements of social organization and situated processes of information searching through the operation of which it comes about that a particular type of knowledge or item of information emerges as significant for the situation at hand. For example, one and the same "case," say an instance of drug addiction, may be embedded in rather different structures of relevance, depending on whether it is seen in a medical or legal context, or in the context of a personal scandal with political ramifications. In each of these structures different modes of defining the "problem" lead to quests for different types of knowledge. Or one might think, as another example, of public debate in which a particular theme emerges which is recognized as the subject of the debate. Indeed, one may think of such structures of relevance as having formal attributes in analogy to the structures of a methodic inquiry. For example, in the courtroom in the course of a trial there emerge certain technical and substantive questions as the strategic ones on which the outcome hinges. There is a formal, symbolically defined structure to the proceedings which much shapes what ultimately is deemed relevant and what is not.

Structures of relevance have a social and cultural aspect that interact. The cultural aspect is emphasized, for example, in the symbolically defined rules of inquiry, the coherence of the arguments by means of which knowledge search is pursued, and by the logic of the domain. The relevance of items of knowledge to a legal debate is determined through such coherent arguments and the rules of the law. At the same time, structures of relevance have a social and organizational aspect in the form of socially structured settings for inquiry and knowledge search. For example, many studies find that interpersonal networks of communication play a major role in the diffusion of knowledge. The linkage between these social networks and culturally defined knowledge criteria is apparent in the finding that networks as such are not sufficient to determine the credibility of diffused knowledge. For example, Elihu Katz (1961:72–82) finds that interpersonal communications networks describe the pathways through which knowledge diffusion travels; however, early adopters of the innovation in question act rather cautiously, trying to assess it in terms of their own reality tests. Late adopters, seeing the experience of the early adopters, proceed more audaciously on this reassurance. Another example can be found in the work of William J. Paisly (1968:1–30), who surveyed the behavior of scientists and technologists in gathering information. Paisely found that interpersonal networks of colleagues and work teams served as a significant information source for technological workers. On the other hand, researchers relied more heavily on formal sources in the professional literature. The acceptance of information thus appeared more related to the technical, professional, and intellectual quality of a channel of communication than to its mere accessibility. It is this combination of social, structurally provided channels with culturally defined criteria that leads to the assessed trustworthiness of a source in the context of a structure of relevance.

Few such structures of relevance have become routinized and institutionalized to the extent that they operate with a certain automatism. However, there are some examples: bibliographical services, computerized data archives, and the like have attempted to find mechanical or electronic methods for delivering information that matches the knowledge search behavior of the potential user. Through such work, information science has made progress in understanding structures of relevance. However, the emphasis on technical need has often limited the necessary awareness of the social embeddedness of such enterprise.

We again need to remind ourselves of the embeddedness of these structures in the context of common sense. It is the break in routine that leads to a reorientation of practice that may initiate a deliberate knowledge search. Common sense connects with specialized search procedures by providing guidance in gaining access to these special domains. In spite of the high segmentation and differentiation of the American knowledge system, the common sense rules of conduct encourage a broad knowledge search. This is quite different from the traditional pattern in many European societies where structures of relevance were more formally tied to the system of occupations and statuses.

The basic structures of relevance guide knowledge search activities in everyday life. In contemporary society these include generalized information resources such as libraries and reference works. More specifically, such structures are linked to existing epistemic communities, be they professions or disciplines, and their gatekeepers. One may think of the family physician as gatekeeper for an individual's relevance structure concerning the health area. Yet, the example immediately points up that the institutionalized, official pattern is surrounded by alternative channels of knowledge search. Persons with experience in the immediately accessible social network and professionally unsanctioned health services provide alternative routes in which reality tests diverge radically from those based on empirical science. Basic relevance structures, then, are those established by institutional channels of expertise and considered trustworthy in terms of well-anchored beliefs.

Yet the impression of stability that this description gives may well be misleading. Secondary relevance structures, superseding institutional patterns, arise through knowledge change, fashion, and the formation of public issues. Fashion certainly plays an important role in that the appropriate themes and topics one talks about emerge as distinctly status-related phenomena. The upsurge of interest in acupuncture, after the emergence of the People's Republic of China as a major power in the world scene and President Nixon's trip, can serve as an example. Acupuncture had been known in American medicine for a very long period. However, spectacular reports raised it to a topic of fashion. This phenomenon undoubtedly contributed to the attempt to incorporate acupuncture into scientific medicine (Rosenberg, 1977).

The availability of relevance structures alternative to those officially sanctioned is not only influenced by fashion. The pressing needs generated by a problem may not be answered by available knowledge resources so that a search for alternatives occurs. This is influenced by the distribution of trust in knowledge resources and their gatekeepers. It is not uncommon that patterns of trust and distrust in the larger social structure, for example between ethnic, racial, or religious groups, confine knowledge research patterns to the ingroup and its subculture.

The practice of science-based medicine in many parts of the world has been viewed with less than complete trust and acceptance because of its linkage to western civilization and power. The very continuation of Chinese traditional medicine in the context of a rapidly modernizing city like Hong Kong illustrates this point.

Those patterns in relevance structure may be deeply affected by the emergence of political issues and cleavages. In a different sense, the construction of an issue in the political arena typically involves deliberate attempts to influence emergent relevance structures among concerned audiences.

The deliberate restructuring of pathways into the knowledge use system has been an important aspect of modernization, social development, and policy for some time. For example, the introduction of modern agricultural production techniques through the extension system of universities and the·county agent network in the United States was a successful attempt to create linkages and knowledge awareness where they did not exist before. Traditional relevance structures of farmers, who used the role-embedded knowledge of family and community patterns, were supplemented and in many instances superseded by deliberately designed channels. Similar efforts have occurred in the educational system. Even more spectacular has been the emergence of research and development complexes in industries that attempt to manage and regulate knowledge flow from the side of production as well as from the side of use.

Models for Knowledge Application and Their Social Forms

We have discussed certain cultural models for knowledge application, such as the Kantian, the Marxian, and other conceptions. Our examples were taken from the history of social thought, dealing with influential philosophical ideas. We now turn to certain cultural models in contemporary, post-modern society. We use the term *model* in the broad sense of a cultural pattern that articulates learnable conceptions for action. Thus, every major institutionalized pattern can also be said to represent a cultural model. Our examples will include certain deliberately constructed designs used in attempts to manage knowledge systems for their components. This encompasses certain social science descriptions and prescriptions; here the term model is conventionally used simply as a name for some fairly formal descriptive scheme that explicates key relations in a phenomenon under discussion. We subsume this more specific social scientific usage under our broad one as a special case because some social science models also have prescriptive implications; they were devised for the purpose of intervening in some fashion in the knowledge use process. We hope to make clear that there is a curious limitation in the instrumental designs proposed by social scientists that can be overcome by considering broader conceptions of cultural models for knowledge use. These have emerged historically without having been intentionally devised instrumentalities of self-conscious, policy-making authorities. In the following paragraphs, we present some of the instrumental conceptions for knowledge use recently discussed in the United States.

Knowledge Production and Use Systems.   Under a variety of names the idea of integrated knowledge production and use systems that could be centrally managed has achieved fairly wide currency in recent years. For example, the National Institute of Education is attempting to construct a national system of research and development organizations that would be systematically linked to programs in knowledge implementation. The Center for the Interdisciplinary Study of Science and Technology at Northwestern University, under the directorship of Michael Radnor, has for some time devoted its efforts to the systematic study of research development and implementation systems. The systems perspective is important here. It seems to be applied primarily in terms of the notion of deliberately designed and centrally steered management systems. In a report (Radnor et al., 1976) to the National Institute of Education, the Center for the Interdisciplinary Study of Science and Technology writes, for example,

> . . . our approach grows out of and builds on a systems perspective, with NIE's mission being viewed in terms of its impact, as an integral part of the system, on the educational R/D&I system's health, functioning and outputs. Further, growing out of this systems perspective, and as is inherent in our general analytical method, we engage in a broader, more systematic analysis of R/D&I functions and the range of conditions affecting the system. . . . Thus the essence of the issue as we see it is: how does NIE achieve its purposes through procurements and other agency actions, taken in concert with and as part of the field?

Whether these constructions are called knowledge production and use systems or research development and implementation systems, they share the premise that knowledge production should be managed in conjunction with knowledge storage, distribution, and use. This particular cultural model is an outgrowth specifically of post-modern society, which generates such comprehensive and ambitious policy objectives. At the same time, there is no instance in which central control of such an institutional sector has been complete. Indeed, in many instances such attempts have been less than decisive with regard to knowledge development and use because such deliberate policy oriented efforts are easily swamped by larger social and cultural forces.

A recent summary of certain models for knowledge use has been compiled by the Human Interaction Research Institute (1976) in collaboration with the National Institute of Mental Health. This volume does provide a systematic appraisal of the literature in the field of knowledge utilization. The authors conceive of the area as a field of study that develops insight, both for knowledge producers and users, into the patterns of knowledge production, distribution, and use. They attempt to identify those factors that impede knowledge use and adoption in order to generate strategies for more timely knowledge utilization (1976:2). We will confine our discussion to six widely quoted models that they describe (1976:66ff). These six models emphasize one or more phases in the continuum from knowledge production to use; they are thus not entirely and sharply distinct from each other. The "research, development, and diffusion model" is based on the assumption that knowledge production is the essential process. It assumes a linear relationship between production and knowledge use, since the potential knowledge user is conceived of as a relatively passive target audience, likely to accept an innovation if it is presented in a rational manner.

> It calls for a rational sequence of activities from research to development to packaging before dissemination takes place. It assumes large-scale planning, and requires a division of labor and a separation of roles and function. Evaluation is particularly emphasized in this model, in which there is a high initial development cost and which anticipates a high pay-off in terms of the quantity and quality of long-range benefit through its capacity to reach a mass audience.*
>
> *(Human Interaction Research Institute, 1976)*

Clearly, a diffusion of knowledge from the source of production to the point of use through organized attempts at distribution is assumed here.

*Reprinted by permission of Edward M. Glaser.

The second model the authors term the "social interaction model." This conception recognizes that a user may hold a variety of positions in the network of communication. It emphasizes psychological and attitudinal factors that influence the acceptance of communications, such as reference group behavior. The conception underscores the intricacy of social relationships and social subsystems involved in knowledge diffusion.

The third, the "problem-solving model," differs from the first two in that it begins with the user's need. It takes a perspective similar to that discussed under structures of relevance. However, the problem-solving model does have a prescriptive aspect (Ibid.).

> The outside helper, or change agent, in this model is largely nondirective, mainly guiding the potential user through his own problem-solving processes and encouraging him to utilize internal resources. The model assumes that self-initiated and directed change has the firmest motivation and hence the best prospect for maintenance.

The fourth conception is that of "planned change." This model assumes that (Ibid.):

> Change occurs through a consciously controlled, sequential, and continuous process of data generation, planning and implementation.

It is a recipe for organizational change in which information is considered relevant only if it leads directly to action.

Fifth, there is the "action research" model. This emphasizes the conduct of research on the site of knowledge utilization, within the using organization (Ibid.):

> The type of research and its methodology are influenced by its concurrent conduct with the ongoing activity of the organization. The results of the research, while primarily intended for the organization itself, may prove useful to others and contribute to behavioral science itself.

This conception is similar to the one emphasized in our earlier discussion of Robert Chin's work.

Finally, there is the "problem-solving dialogue or linkage model." This rather complex conception emphasizes four components: the user system, the knowledge resource system, a need-processing system, and a solution-processsing system. Here the flow of knowledge and information is treated as a two directional process. This conception has been applied as a scheme for the study of several federal research dissemination agencies.

Intervention and Planned Change. The conceptions of knowledge production and use systems and the six models we have just described are instructive. All of them are based on the desire to intervene planfully in the knowledge utilization process. In this sense we could argue that they themselves are the products of a cultural movement that links professional social science with certain segments of public administration and government. This circumstance may also account for their rather obvious limitation in treating knowledge application processes as basically unidirectional from knowledge production to use. Further, there appears to be a strong managerial concern in many of them, reflected in their primary interest in artificially constructed knowledge use patterns.

However, not all recently articulated models are properly subsumed under this description. The idea of informed intervention in social change as a desirable form of knowledge application has linked professional efforts with the interests of social movements. For example, the conception of action research in which the professional investigator helps to clarify situations for those involved in change efforts seems to be broadly similar to attempts to use research in the context of community change and improvement. Thus, Saul Alinsky's mobilization and participatory organization efforts in communities were much influenced by social scientific conceptions. This suggests that while official strategies for knowledge utilization are important historical phenomena, they must be seen in a context in which the very idea of knowledgeable, and hence effective, intervention has gained much broader appeal. The phenomenon of the professionalization of contemporary movements is closely linked to this trend.

Emergent Patterns Versus Managerial Models. There are cultural models for knowledge use of massive historical and contemporary importance that are only in part supplemented or superseded by deliberately constructed ones. The gradual institutionalization of invention may serve as an example of such an historically emergent pattern. The transformation of invention from an occasional activity of the individual tinkerer to a full-time, organized and managed pursuit in research and development organizations is quite apparent. The *de facto* emergence of organizational experience with research and development complexes clearly preceded the deliberate design of managerial approaches and related conceptions that we have discussed. Distinctive styles of administration and of authority relations have appeared in these organizational patterns. The relatively high level of autonomy in work itself is combined with efforts at coordination and management. The idea of *managing by objective* illustrates this pattern. It has been subjected to scrutiny by Swanson (1979) as a new basis of authority and identity in post-industrial society. There are other such cultural models for knowledge use that have emerged historically without having been deliberately designed. There are patterns of professionalism, policy formation, social movements, and entrepreneurship.

The cultural model of professionalism is probably one of the oldest patterns of explicitly prescribed modes of knowledge utilization. The predominant pattern even now is that of the professional linking a client to a body of knowledge via a role-embedded expertise. The explicitly knowledge-based professional who practices predominantly in terms of technical, publicly accessible scientific knowledge even now is relatively rare. It is the reliance on role-embedded expertise and the relatively localized control over patterns of professional practice that guaranteed a high level of professional autonomy and authority. Yet, while the pattern of professionalism is selectively changing towards a more explicitly knowledge-based form, the core phenomenon remains the professional as an expert who absorbs uncertainty for clients. Such risk absorption is expressed in the notion that a "client puts oneself into the hands" of the professional. This is done, however, on limited terms. The professional necessarily exercised control over the situations by defining them in terms of personal criteria, rather than that of the client. It is clear that every major profession can be seen as a link between knowledge use and knowledge production, with an elaborate structure of schooling, regulation, and standard setting involved in the system.

Excursus: On Professionalism, Ideologies, and Knowledge Use.
It is precisely the claim of professional occupations that they can
absorb risk and uncertainty, coupled with actual knowledge deficits,
that makes professional arenas fertile grounds for the construction of
ideological belief systems. We have already referred to Geertz's
(1964) conception of ideology, which he applied to examine the
interplay of political processes and cultural symbols in a nation
undergoing rapid change. However, his conception is fruitful also for
the understanding of the relation of professionalism, specialized
occupational ideologies, and the demands of knowledge use. Specifically, the ideologies that are associated with professional subcultures
represent consensually validated belief systems about problematic or
incomprehensible aspects of reality on which the practitioner must
act in order to live up to claims for competence and influence. These
ideologies serve as guidelines and rationales for purposeful professional action by providing shared meanings and interpretations of
incompletely understood situations.

On the basis of this conception of ideology, it is possible to
isolate the particular attributes that make professional arenas strategic contexts for contemporary ideologies. More specifically, what
distinctive characteristics of professional fields are responsible for
generating ideologies? They turn out to involve issues of knowledge
application.

Most attempts to define profession and distinguish professional fields from the rest of the occupational structure include some
references to a "systematic body of theory" (Greenwood, 1957:45),
"abstract knowledge" (Goode, 1957:194), "intellectual content"
(Kornhauser, 1962:1, 8-11), or "systematic knowledge or doctrine
acquired only through long prescribed training" (Wilensky, 1964:
138). Thus, a crucial characteristic distinguishing professional
behavior from that found in nonprofessional occupations is that it is
(presumably) based on and guided by systematic abstract knowledge
about the subject matter of professional concern. What is frequently
overlooked, however, is that this body of knowledge is generally
incomplete and provisional. There are few, if any, professional fields
in which the behavior of practitioners can be completely and definitively guided by a body of available knowledge. That is, there are few
arenas in which the subject matter of professional concern can be
completely understood, predicted, and controlled on the basis of the

body of knowledge available to professional practitioners. It is precisely this confluence of a normative expectation that practitioners' behavior will be based on and guided by the extant body of systematic abstract knowledge within a field, and an awareness that this knowledge is inadequate for understanding, predicting, and controlling the subject matter of professional concern, that sets the stage for the rise of competing ideological belief systems in professional arenas.

The significance of ideologies in a professional arena is inversely related to the extent to which the phenomena or situations that practitioners must deal with have been completely and definitively understood. When complete understanding of the subject matter is obtained, competing ideologies can be replaced by broad agreement as to the appropriate general approaches and the more specialized techniques for dealing with any problem requiring professional attention. On the basis of these considerations, it is possible to suggest some characteristics of professional fields that are highly likely to generate ideologies in response to the inadequacy of other sources for determining professional behavior. They are:

(a) Fields that are relatively new or recent and have undergone rapid growth in size, scope, social demand, or significance. These conditions maximize social pressures for professional action while at the same time the practitioner is unfamiliar with, and uncertain about, the phenomena toward which action is demanded. The same conditions minimize the likelihood that the profession's intellectual resources are adequate for dealing with these phenomena.

(b) Fields in which the application of empirically validated knowledge to concrete social problems depends on personal, subjective-intuitive, particularistic attributes of the practitioner. Where extrascientific considerations inhere in the application of scientific knowledge to social affairs and where practitioners must rely, to an important extent, on the idiosyncratic art of professional practice, the stage is set for divergent interpretations and competing ideologies.

(c) Fields in which moral and ethical considerations surround both the subject matter and the ends of professional action (Bendix, 1964:296). These considerations are usually phrased in terms of professional issues over social values—rather than in explicitly philosophical terms.

The relationship between characteristics such as these and the importance of ideologies as alternative guidelines and rationales for professional behavior is nowhere better illustrated than in the mental health field.

The past few decades have seen increasing attention devoted to the subject of mental health and illness, resulting in a vast, and largely unmet, public demand for professional treatment services. Nevertheless, the nature, cause, and treatment of mental illness are as yet only partially understood and have been the subject of much public and professional controversy. As a result, professional practitioners have been forced to take therapeutic action toward mental health problems in the absence of certainty as to the most effective preventive and therapeutic approaches—or even the long-term consequences of some approaches. Moreover, many therapeutic approaches touch on fundamental, moral and ethical issues not yet resolved in the larger society.

These conditions expose mental health practitioners to a set of patterned professional strains or tensions that are associated with their central function, namely treating mental illness. These patterned strains associated with the core professional roles and functions of mental health practitioners have generated alternative ideological belief systems concerning the treatment of mental illness.

The ideologies held by mental health professionals, then, are shared belief systems that guide and justify purposeful therapeutic actions in the absence of complete, scientific knowledge about the nature, prevention, and, most importantly, the treatment of mental illness. Thus, these ideologies represent a response, which is supported by the consensual validation of other professional practitioners, to the strains associated with professional responsibility for treating mental illness on the basis of incomplete and provisional information.

POLICY, MOVEMENTS, MOBILIZATION

Another broad institutionalized pattern for knowledge use can be found in conceptions of policy formation. Of these, the notion of reflective incrementalism, in which a continuous cycle between decision, action, and assessment through research plays an important role, has recently become particularly prominent as an explicit value in democratic policy conduct. Although policy formation has followed quite different rationales in other contexts and times, it was always closely related to problems of knowledge search and utilization. Indeed, conceptions of rational behavior, emphasizing empirical inductive or rational deductive approaches respectively, probably played a major role in shaping the policy-forming styles of regimes. The emergence of the modern nation-state can be shown to be linked to knowledge production efforts of Herculean scope. For example, the proliferation of social accounting, indicators, and measurements for the purposes of monitoring sociopolitical and economic processes accompanied the growth of rational administration and central bureaucracies. Hence, it appears that whatever the pattern of policy formation, it will be intimately connected with the institutional structure of the larger knowledge systems. Certainly, the conflicts about governmental or other efforts to restrict the availability of policy related information have profound effects on the knowledge system at large.

We have further listed social movements as patterns of knowledge utilization. By movements we always mean some deliberate mobilization for or against change; they are always based on conceptions of alternative possibilities and ideas of appropriate pathways to desired future states of affairs. Knowledge, particularly social knowledge, has played a demonstrably large part in shaping social movements; it enters into the construction of ideological belief systems as well as the devising of tactics and strategy. The patterning of movements has been profoundly influenced by the availability of such knowledge. But the relation can be turned around. It is necessary to recognize that the structure of the knowledge system itself not only influences social movements and their possible directions, but that it is in turn powerfully influenced by them. It can be shown that movement patterns have historically changed from mobilizations largely confined by tribal, ethnic, or otherwise ascribed boundaries to more inclusive mobilization attempts. The idea of revolutionary social movements that deliberately substitute newly designed social structures for existing ones is historically a fairly late development. Even later, we would argue, is the emergence of movements of psychological well-being and the transformation of individual identity in explicitly knowledge-based terms. These movement phenomena are predicated on what is taken as knowledge about possibilities for change.

Social movements create novel structures of relevance by generating issues, mobilizing resources, and delineating domains of conflict. Through such impacts, the knowledge system is profoundly affected. To illustrate this point, consider the various socialist movements and their effects not only on political life but also in social science, specifically with respect to the debates concerning social knowledge. More recently, we find that cultural movements for changing personal identity claims go hand-in-hand with attempts to alter the knowledge system in significant ways. The introduction of women's studies accompanied the movement to change the sense of identity and status of women in contemporary society. Such powerful mobilizations link the knowledge system with the most basic currents of social and cultural change. In post-modern societies, governmentally designed or otherwise artificially created knowledge use systems receive much of their effectiveness from the extent to which they supplement such major currents of social change rather than contradict them.

Entrepreneurship is our final illustration of a pattern for knowledge use that emerged historically and remains a powerful force. The orientation to opportunity and bridging of gaps between different frames of reference is a characteristic of the purposeful use of resources in the hands of entrepreneurs. Our examples of innovation have shown the significance of technical entrepreneurship in knowledge use. In spite of refined management techniques, training methods, and whatever other systems are invented, there remains no substitute for the ingredient of leadership inherent in the effective entrepreneur. This observation should serve to remind us that the knowledge system is a living component of social life, influencing its shape as it is shaped by it.

This point concludes an overview of processes of knowledge application. Chapter 9 will examine the role of the knowledge system and sociocultural change, particularly emergent relations between authority and identity in post-modern society.

PART
III
Bibliographic Notes

The three chapters that form Part III of the book attempt to delineate some of the major features of the post-modern knowledge system and raise some of the central issues that this system poses for contemporary society. A central focus of these chapters is on the strategic significance of several distinct types of epistemic communities for the production, organization/distribution/storage, and application of specialized knowledge in contemporary society. Specifically, we focus on the emergence and differentiation of specialized epistemic communities into the modern scientific disciplines, practicing/consulting professions, and cultural movements. As a result of the recent proliferation of valuable contributions to the sociological literature on science, the professions, and social movements, it is important to reiterate our earlier statement that the works included in these bibliographic notes are mentioned without any pretense of comprehensiveness or completeness. Instead, our emphasis here, as in previous bibliographic notes, is on those works that were decisive for us and that we believe others will find equally stimulating and suggestive.

Like so many sociologists, we have been much influenced by Thomas Kuhn's *The Structure of Scientific Revolutions* (1962 and 1970). This book is particularly congenial to sociologists because of its stress on the role and mechanisms of consensual validation as one grounding of scientific knowledge and its emphasis on scientific and intellectual discontinuities. Post-Kuhnian sociology of science has emphasized the notion that the validity of truth-claims is mediated through the consensus of some scholarly (epistemic) community. Thus, what any scientific or professional community takes for established knowledge rests on the consensus of those it defines as competent and full members of its community. In short, the consensus of the knowledgeable is a necessary mediating condition for believing truth-claims to be valid; such consensus is not, however, a sufficient condition—since consensus may be achieved in "illegitimate" ways. Although advanced by Kuhn, the views of science as the work of a community in the sociological sense was first formulated by Michael Polanyi in works such as *The Logic of Liberty* (1951). In its more extreme form, the Kuhnian model leads to the refusal to distinguish between science as a body of knowledge, science as what scientists do, and science as a social institution. For an example of this, see John Zinman's *Public Knowledge: The Social Dimension of Science* (New York: Cambridge University Press, 1968). For a balanced sociological treatment see Warren O. Hagstrom, *The Scientific Community* (1965).

A different approach that is more limited in scope, but powerful in its explanatory capability can be found in the institutional sociology of scientific activity entitled *The Scientists's Role in Society: A Comparative Study* by Joseph Ben-David (1971). This remarkable book analyzes the conditions that determined the level of scientific activity and shaped the roles in careers of scientists and the organization of science in different countries at different times.

Another significant body of scholarly work has emerged around questions concerning scientific and technological growth and differentiation. On these and related subjects the work of Derek J. de Solla Price has been particularly influential, especially his *Science Since Babylon* (1961) and his subsequent volume, *Little Science, Big Science* (1963). No selection of readings in any of the areas that touch on the sociology of science would be complete without mention of Robert K. Merton's pioneering study *Science, Technology and Society in Seventeenth Century England* (1938). We further rely on Robert K. Merton's *The Sociology of Science* (1973) edited by Norman Storer. Perhaps the most notable attempt to combine an interactional study of scientific activity with an appreciation of the conceptual and logical structure of science—as reflected in its paradigmatic status—is Diana Crane's analysis of *Invisible Colleges* (1972). And among the many recent worthwhile collections of articles and essays dealing with science and knowledge, two issues of *Daedalus*, the *Journal of the American Academy of Arts and Sciences*, warrant explicit mention: the Spring, 1973, issue entitled "The Search for Knowledge" and the Summer, 1974, issue entitled "Science and Its Public: The Changing Relationship." Finally, one contemporary sociologist who has produced significant contributions both to the sociology of science and to the sociological literature on professions is Joseph Ben-David; in addition to his book on science and numerous specific articles dealing with structural changes in universities and the growth of science, he has also produced a number of works on specific aspects of professional development. Thus, Ben-David (1960a) has described "Scientific Productivity and Academic Organization in Nineteenth Century Medicine" as well as the broader issue of "Roles and Innovation in Medicine" (1960b).

The general body of literature we have found most useful as background material for addressing questions of the organization, distribution/dissemination and storage/retrieval of knowledge in post-modern society is usually subsumed under the label of communications research. Discussions of the philosophical foundations for a communication theory of society, as well as some useful basic data, are to be found in I. de Sola Pool et al., eds. *Handbook of Communication* (1973) and in W. Schramm, ed. *The Science of Human Communication* (1963). Another useful collection has been edited by Denis McQuail entitled *Sociology of Mass Communications* (1972). A number of recent critiques of the concentration of the mass media bear directly on the issue of the distribution and dissemination of various kinds of knowledge in post-modern society. These deal with the growing concentration of information systems in the private sector and their increasing fusion with government in western societies, and, especially, the United States. Unfortunately, few of these critiques articulate what kinds of workable, democratic *alternative* new systems they have in mind. That is, the full meaning of diverse recent critiques of media concentration cannot be appraised seriously unless there is a clear specification of the alternative arrangements sought or condoned. One work that begins to move in that direction is Claus Mueller's *The Politics of Communication* (1973); another such piece is Hans Magnus Enzensberger's *The Consciousness Industry* (1974). Both of these works owe a considerable debt to Hugh D. Duncan's creative landmark in this area, *Communication and Social Order* (1962). Finally, useful contributions in the Lazarsfeld and Merton tradition of communications research at Columbia University include Herbert Menzel's article in *Public Opinion Quarterly* (1972) entitled "Quasi-Mass Communication" and Richard Maisel's *Information Technology* (1976).

We already have acknowledged our indebtedness to Eliot Friedson's book, *Profession of Medicine* (1970). An almost equally useful work is the collection of essays that Freidson edited entitled *The Professions and their Prospects* (1973). This volume contains, among many other useful essays, the intriguing paper by Albert L. Mok entitled "Professional Innovation in Post-Industrial Society" (1973). It would be inexcusable to omit reference to Everett C. Hughes' classic, *Men and Their Work* (1958). And any bibliographic suggestions for further reading on the professions would be incomplete without two papers that are rapidly becoming modern landmarks in the sociology of the professions: Harold Wilensky's "The Professionalization of Everyone?" (1964) and Marie R. Haug's article, "The Deprofessionalization of Everyone?" (1975). These two conflicting pieces are discussed more extensively in the Postscript to the Appendix on the Professions at the end of this volume. Finally, it would be impermissible to round out the notes for Part III without mentioning E. M. Rogers' appraisal of the sociological literature on *Diffusion of Innovations* (1962).

PART
IV

Concluding Remarks

Chapter 9 concludes the book with some speculations on sociocultural change in the light of our analyses of knowledge application and the knowledge system. These observations bring the discussion back to some of the general themes raised at the beginning of the book, but treat them from a different perspective. A central premise of this chapter is that personal and collective identity constitutes a pervasive concern in post-modern cultural systems. We discuss how this preoccupation forms a focus for diverse attempts to apply specialized social scientific knowledge to the lives of individuals. These attempts range from consciousness-raising groups in cultural movements to the triumph of the therapeutic *in the sensitivity training-encounter group movement as well as in the educational and judicial systems. The pervasiveness of therapeutic models of identity and personal growth or change is discussed and interpreted in relation to questions of accountability and autonomy.*

9

Speculations on Sociocultural Change

INTRODUCTION

In this chapter, we attempt to delineate some broader connections between the themes treated in this book and major foci of
current social and cultural change. This brings us back to some of the
general considerations and issues raised at the beginning of the book.
Our central concern remains the knowledge system that we have
examined in terms of basic reality construction processes as well as
conceptually distinguished phases involving knowledge production,
organization, distribution, storage/retrieval, and application or use.
But these analytically specialized interests are embedded in broader
concerns about the shape, direction, and meaning of emerging social
and cultural forms. The task of systematically describing the cultural
templates and domains that structure the emergence of post-modern,
knowledge-based societies goes well beyond the scope of this book.
Moreover, our emphasis on the diversity of embedding contexts that
structure knowledge-related processes left many broader issues and
phenomena that could not be touched on. In this final chapter we
grant ourselves even greater license than before and take a more
global as well as inevitably speculative stance.

Interpretations and forecasts concerning emerging sociocultural structures and trends are precariously risky and fragile. Questions about the impact of the contemporary knowledge system on
broad processes of sociocultural change and vice versa lead us beyond

the reassuring boundaries of what is reliably known. Moreover, there is simply no single compelling pattern or trend that can be readily seized, described, and interpreted. Instead, the contemporary social and cultural scene is like a colorful, multidimensional kaleidoscope, changing and shifting rapidly, frequently in largely unforeseen fashion. In addition, many of the more unambiguous sociocultural developments and trends are accompanied by equally prominent countertrends that generate complex dialectical strains and processes of change. These considerations require that we approach the task of this speculative concluding chapter informed with an appreciation of the wide range of the possible, sharpened by skepticism about recent forecasts and visions of the future.

Recent visions of the future have been of two sorts. One kind has foreseen a period of unparalleled freedom, with people possessing the autonomy and leisure they have sought for ages, based on the increasing spread of systematic knowledge and a technologically produced abundance. The economy of abundance in these visions of a post-modern future would care for many of humanity's needs and the omnipotent state would not exact a loss of freedom in selfhood. Blessed with unprecedented amounts of leisure time, freed from the bonds of necessity, and provided access to new realms of knowledge, people would attain new command over themselves and the environment. The other version sees people as inevitably enslaved by an increasingly monolithic, technocratic state, surrendering to abstract, complex, authoritarian institutions that destroy freedom and selfhood. In this version, humanity will be controlled by unseen and unseeing hands that dictate every action and thought. People will be programmed, either genetically or through thought control, free only to obey.

Regardless of which vision of the future proves to be more accurate, both are grounded in expectations concerning the capabilities of systematic knowledge application and technological development, not only on the behavior of humans and/or organizations. And confronted with such extreme contrasts in prevalent images of the future, it seems implausible to maintain even that the truth will lie somewhere in the middle. The capacity of social theorists to foresee or forecast the future on such a global scale has been extremely limited at best. We have suggested several reasons for being skeptical about the possibilities of overcoming the limitations and constraints responsible for this social scientific myopia. For example, we have alluded to the unavoidable embeddedness of observers who can never completely transcend the limitations that particular historical locations place on a social perspective or interpretive frame of reference. Other constraints and limitations on a frame of reference result from any particular focus of analytic attention. Thus, we must exercise special caution to avoid implying that post-modern society can be understood from the vantage point of an analysis of the knowledge system exclusively.

Finally, social science has a dismal record when it comes to sensing significant sociocultural developments ahead of time. For example, most attempts to explain why feminism re-emerged precisely when it did in the mid-1960s had to conclude with Ferris (1971:1) that "from the close perspective of 1970, events of the past decade provide evidence of no compelling cause of the rise of the new feminist movement." His examination of time-series data over the previous twenty years did not reveal any significant changes in traditionally sensitive sociological variables that could account for the emergence of a women's movement at the time it was created. Indeed, it is clear that the behavioral sciences have not responded quickly enough to the challenge of emerging trends and movements after they have burst on the scene.

For example, Joseph Schumpeter published an essay in the 1929 alumni magazine of the University of Bonn in which he made an earnest scholarly effort to describe the direction of social change in the German Reich. It is not a casual piece, but a foray well-informed by the conceptual apparatus of the social sciences of his time. This essay documents one of the remarkable failures at pre-science on the part of social science, so painfully obvious in the light of the Hitler tyranny that arose in Germany only a few years later. In the summary of his essay, Schumpeter wrote (translation ours):

Let me only point to one result of our overview: it is the very great and possibly even increasing stability of our social conditions. For the foreseeable future there must be an overwhelming majority against any extreme force no matter in which direction it would go. That means politically that there can be no govern-

ment whose opposition would not be almost as strong as its support—except if the government were to include such differently
oriented elements that it could act only in the narrowest possible
frame even without opposition. And there is a similar cultural
consequence. In no sense, in no area, in no direction can we expect strong beginnings, upturns, or catastrophes. This is a terrible
situation for people with the burning desire to create and innovate—but for achievements of another kind maybe it is not a
bad framework.

(Schumpeter, 1929)

Since Schumpeter's time, there has been improvement in the
methods and concepts of social science, but its record in discerning
the shape of the future remains in doubt.

Another point must be made. Much of this book has been
written in general sociological terms, yet most of its illustrations are
taken from the experience of the United States. It is difficult to
know what is unique about the American experience in moving into
a new social and cultural era, and what features are more general.
Even if there are similar challenges, there will be different responses
in different cultures. There are broad similarities in the emerging
knowledge systems of all complex societies. Indeed, science itself
has, probably more than any other domain of social life, achieved a
world-wide community of knowledge exchange—in spite of all its
political and economic limitations. Patterns of knowledge use, however, are so specifically geared into the social and cultural fabric of a
civilization that, in spite of all commonalities of the utilitarian sort,
they will reflect diversity. But the problem of comparisons is beyond
our scope; its intrinsic difficulties are compounded by the complex
patterns of positive and negative emulation of the United States
around the world. When, in this chapter, we primarily address the
United States' case, we remain aware that, in many ways, it probably
is unique. Cultural models and religiously grounded value configurations have played a special role in the United States since its founding. The polarization and ossification of class structure, so characteristic of many European nations, has not occurred here. Instead, the
United States has historically shown a remarkable capacity for continuity in political and economic development.

In a recent essay, Tiryakian (1975) has argued that most of the classical (exempli gratia, Marxian) models of European social theory do not fit the American case. Tiryakian writes (1975:8):

> In European imagery, "the revolution" is either a glorious achievement of the past, heralded as the creation of modern society, upon which the ship of state is anchored and which disguises the absence of real social development—or else it is placed in the eschatological vision of the future. In the United States, it seems to me, the revolution is not so much a sacred point of history as the continuous historical process itself, given what I see as a dominant value orientation of commitment to actualize change, not only in the technological sphere but also in the multifaceted social sphere.

Tiryakian finds Durkheimian models to be wanting as well. For example, following this model one would expect an increasingly paralyzing crisis of American morality as religious tradition and the institutions on which it is based lose their significance.

> Yet I would argue that were all the evidence available, we might be able to demonstrate that in this century, with sharply increasing social heterogeneity, moral accountability has become more and more serious in the public life of the United States—in such matters as the regulation of corporations, race relations, domestic life, environmental affairs, and the behavior of government officials. The mighty stream of moral accountability is always at the gate of the American forum!
>
> *(Ibid.:13)*

The reason that Tiryakian rejects both the Marxian and Durkheimian conceptions, but tentatively accepts Weber's, stems from his conviction that Puritan social theory created central features of the way in which Americans perceive the relation between self and organized society. Tiryakian sees a transformation of religiously grounded ideas of the community into secular ones that have considerable force. The American conception of the community "is an organic reality but one resting on freely-arrived-at contractual relationships between free agents—the community is not a superhuman entity" (Ibid.:25).

Tiryakian's argument underscores the lack of fit between the historical reality of the United States and major portions of the theoretical legacy of social science. His conception of the centrality of individualistic value themes and a basic code differing from most known European patterns deserves attention. This American code should also manifest itself in the structures and consequences of the knowledge system. Yet, even in this regard, there are trends and counter-trends and the future remains uncertain. We now turn to an examination of the expansion of symbolism and the specialization of cultural domains.

THE EXPANSION OF SYMBOLISM AND SURFEITS OF INTERPRETATION

Even though the configuration of post-modern culture is still emerging, it is clear that it involves an enormous expansion of modes and media of symbolization. This is obviously true in relation to the expansion of knowledge itself, but it can also be observed in the creation of novel and distinctive symbol domains in art, literature, philosophy, and logic. It is possible to describe this cultural specialization as involving the development of increasingly distinctive languages, systems of notation, and expression: exempli gratia, the expansion of logic from the age-old heritage of Aristotelian thought to the current proliferation of logics. The expansion of mathematics may serve as another, similar example. But this pattern is not confined to technical fields of precise analysis. With some simplification we can say that the proliferation of styles in art parallels this phenomenon in expressive culture.

One important consequence of this diversity and richness in the means of symbolization is the existence of many alternatives and choices in representation. This characteristic of emerging cultural configurations is connected with the existence of a multiplicity of interpretative possibilities for almost any situation. We might refer to this phenomenon as *surfeits of interpretation.* By this we mean that any given situation is capable of many alternative interpretations so that choices among them, not just within them, need to be made.

The multiplication of media may be even more significant than that of symbolism. The manifold means of communication, exchange, and representation that have developed make so much more information potentially available that alternatives of interpretation occur with increasing frequency. In older sociocultural patterns, one important source of implicit social regulation was the absence of alternatives for interpreting most situations. In the post-modern pattern, on the other hand, communications experts and other knowledge distributors see it as their task to enter a *market* of interpretation and offer alternative positions to specified publics. Since in most cases a premium is to be won by having one's interpretations accepted, this is quite a competitive market. It has the obvious consequence of problematizing many domains of social life previously thought of as unproblematic or taken for granted. The proliferation of cultural movements centering on life styles and patterns of conduct, even on identities, is one result of this expansion of symbolism and media themselves.

Surfeits of interpretative possibilities increase the domain of freedom for action and of expressive culture production on the part of specialists of communication but, at the same time, they impose considerable burdens on actors. Few situations now seem so inherently clear that their meaningful interpretation is beyond all doubt. Much more prevalent are situations in which alternative models of interpretation vie with each other and contribute to the problem of uncertainty and uncertainty management.

With some simplification, one might argue that earlier world views provided images of society and the world as fixed structures giving reality a reassuring unity and simplicity. Even the notion of evolution toward an already visible new society, as it occurred in the idea of progress, while not limited to fixed structures, provided a limitation on interpretative alternatives. Such guide posts no longer are readily available. Indeed, the rather extreme oscillation in images of the future between great optimism and deep pessimism clearly illustrates this point.

In view of these developments, contemporary intellectuals are oriented to sets of events, rather than to fixed and immutable structures or to historical processes, the direction of which could be taken as predetermined and unchanging. They have had to become accustomed to orienting themselves to probabilities, rather than to certainties. We had repeated occasion to speak of the structure of post-modern common sense; it appears to us to reflect this orientation to probabilities and methods for their management.

UNCERTAINTY MANAGEMENT

The fact that decisions need to be made in the face of alternative interpretations and choices as well as in the face of uncertainty and, thus, risk is one of the dominant themes of the current cultural situation. Indeed, it currently is taken for granted that certainty, in any final sense, is not attainable in any domain of human knowledge or belief.

Such a focus on alternatives places man as decision maker at the center of attention. Reliance on tradition has become suspect. The search for reliable criteria of judgment, in turn, requires increased reflection, and we have already seen that reflectivity compounds the problem of uncertainty. Indeed, the hope that a knowledge-based society would go about the work of knowing in a proper fashion and replace uncertainty with truth now appears much too simplistic an idea. Our discussion has shown that, virtually inevitably, there is a diversity of epistemic criteria that are not readily translatable into each other. The fact that all such criteria have a discernible relation to an encompassing rationality is quite a different matter. In practice, it is the case that the circumstances of reflectivity, of the multiplicity of perspectives and criteria of judgment, give the problem of action in the face of uncertainty its very special contemporary flavor. It is for these reasons that the emerging era faces the problem of uncertainty and uncertainty management in an historically unique and specific manner.

We are suggesting, then, that the problem of managing uncertainty is a central issue in the current sociocultural configuration. It can be seen clearly in the world of the professions, especially in relations between client and practitioner. It always has been one of the important tasks of the professional to manage the uncertainties of the client, and to draw on knowledge resources as well as on authority and trust in the profession to deal with personal uncertainty problems. The post-modern situation renders this task more significant and also more problematic.

A similar focus on the management of uncertainty exists in the field of policy analysis. Uncertainty management is the phrase we use because we rarely find the simple replacement of ignorance by knowledge, the reduction of uncertainty through firm predictions and explanations. Instead, we find that knowledge application often results in *increasing* uncertainties, either by exposing the unsuspected complexity of a situation or the limitation of available knowledge resources, or through sharper definition of issues. Uncertainty management does include uncertainty reduction through efforts at prediction and control, but it also includes efforts at the absorption of uncertainty into diffuse relations of trust—that are not in themselves necessarily grounded on knowledge. Furthermore, uncertainty management also includes schemes for the diffusion of risk, so that uncertain situations are perceived as less threatening. We mean here the pattern of limiting risks through insurance, limited liabilities, and the like that occurs in major public programs. This includes attempts to diffuse responsibility through efforts to limit the impact of possible dangers.

Since these cultural and social changes are so closely tied to the making of decisions for action and patterns of conduct, they are also fundamental in shaping the relation between individual and collectivity as well as in shaping patterns of knowledge development, distribution, and use. One aspect of this process can be illustrated through a brief examination of changes in the conception of government in the United States.

SOME CHANGES IN THE CONCEPTION OF GOVERNMENT IN THE UNITED STATES

The brief reflections that follow are, of course, no attempt to account for the complexity of changing conceptions of government in American history as such; rather, they illustrate trends in relations between individuals and society as these are affected by aspects of knowledge use. The architects of the Constitution concentrated on rules of representation, the scope and powers of the several branches of government, and the establishment of limits on political action. One might characterize the conception of the republic they fashioned as that of a rational, political state. Social knowledge and measurement, for example through the census, were given a special and important role in this conception of an enlightened republic, making its decisions on the basis of rational discourse. The domain of the political was conceived of as limited and, within its domain, the settlement of disputes and conflicts was thought of as following reasonable debate and adjudication. An enormous domain of private action remained outside governmental relations, and was, indeed, safeguarded from it. Highly significant in this regard, no doubt, was the American separation of church and state.

The growth of industry, and the increase of military and administrative responsibilities, gave the American government many additional functions so that, by the end of the nineteenth century, the conception of government became increasingly that of a regulative agency. Regulatory instruments and agencies were created to extend government into many public activities. Thus the first national regulatory instrument, the Interstate Commerce Commission, was established in 1887, followed three years later by the Sherman Antitrust controls. The basic purpose of this new regulative conception of government was to limit excesses in the private sector, particularly of the new corporate concentrations of enormous wealth and power. Complex organizations had appeared on the scene that, while defined as private, had for many intents and purposes quasi-governmental effects on many people. Even though government began to limit the operation of private enterprise, the separation of public and private and the idea of the competitive private market remained inviolable. Through the regulatory agencies, however, concentrations of functional expertise became established within the governmental structure that gained considerable independence from representative political institutions. Certainly, some modification of the earlier conception of the relation between individual and the polity was involved.

Under this influence in government, the separation of administrative from political orientations was built into the major regulatory agencies. By means of this separation, regulative administration became superior to politics and efficiency replaced representation as the key operational principal of public policy. Regulatory agencies were exempted from the sacred doctrine of the separation of powers and were assigned substantial powers. Larger and more diffuse grants of power were turned over to efficiency experts and new concentrations of functional expertise became established within the structure that gained a good deal of independence from the political institutions.

Beginning with the rise of the conception of government as responsible for many aspects of the welfare of its citizens, particularly in the New Deal era, a new bureaucratic conception began to supersede the regulative administrative one. In this conception, governmental administration was designed not only to correct the deficiencies of the market and limit excesses of concentrations of power, but to supplement it through public enterprise and activism. The regulatory commissions that characterized American government during the administrative period came to be surpassed in importance by the operating bureaus of the federal government during the post-New Deal bureaucratic emphasis. This idea of governmental responsibilities, of course closely tied to the notion of the welfare state, linked administrative and political considerations, uniting them through an emphasis on interest groups. Interest-oriented politics was promoted and resulted in a proliferation of interest and pressure groups with profound effects on the political and administrative process. Conception of goals, involving not only ideals of individuals but ideals for individuals, came to play an important role in this process.

One might illustrate these points by noting the changing role of the courts in relation to government that reflected these different conceptions. In the early eras, the courts tended to tell governments what they must not do; during the much more active bureaucratic era, courts began to instruct governments as to what they are obliged to do in order to satisfy legitimate expectations and legal rights of both citizens and groups. In this context, traditional lines separating public and private domains began to be blurred. The conception of the bureaucratic state is probably still dominant, but there are signals indicating the appearance of a new model for government compatible with a post-modern cultural system.

We already have observed that this emerging cultural model holds governments responsible for knowledge production and use, for the detection and cure of social problems, and even for providing guidelines for action in situations of risky choice. The community mental health movement, so significantly furthered by the federal government, is an example of this last point. Government is seen as one of the main resources available to the individual to limit and absorb risk as well as uncertainty. In this cybernetic model, government functions like a servomechanism, coordinating the needs and capacities of the polity and the economy in order to achieve collective objectives. There is a resulting shift in emphasis from the notion of government as responsible for organizing and producing public activities to government as a distributor of public benefits, as a vast risk absorbing system.

The significance of social insurance in its many forms, of government acting as a collective resource supplementing private efforts at prudence and foresight, can hardly be overestimated. For example, in this conception welfare becomes more or less automated, taking form of a guaranteed income that is adjusted automatically as the income of recipients rises or falls. In this situation, government action is automatically triggered by changes in the economic condition of individuals. Such an emphasis in government differs considerably from the earlier, bureaucratic conception of government in which eligibility and benefits were determined by caseworkers in accord with legislative administrative regulations. Perhaps the clearest indiciation of the shift to a systemic, cybernetic conception of government has occurred in the macroeconomic area, where the refinement of national accounting schemes over the past thirty years has given federal authorities a substantial capability to guide the economy, making quick adjustments as economic conditions change. The expansion of the idea of governmental responsibilities in the direction of risk absorption is much more limited in the United States than in many other welfare states in the world. It also is by no means the only major shift in public notions as to what government ought to be and, as we shall see, seems in the American case to be embedded in a peculiar dialectic. Let us first examine some of the other elements of the emerging cybernetic conception of government in post-modern American society.

The cybernetic conception of governmental responsibility includes the detection and formulation of social problems, so that coordinated interaction between interest groups, intellectuals interpreting the social scene, and professionals emerges around major bureaus of the government. Simultaneously, government has become the most significant source of support in the struggle to expand the frontiers of knowledge, as in the American space program and the increasing government support of basic science. The increasing role of government in safeguarding claims for individual freedom, rights, and dignity, as in the arena of civil rights and liberties, is also relevant here. In fact, it is this turn that may be seen as introducing a complex counter-trend to the dominance of the cybernetic state. The notion that the polity is responsible not merely for public benefits but also for claims to dignity and liberty is a relatively novel idea. On one side, it inevitably triggered an expansion of the governmental apparatus while, on the other side, it reaffirmed value commitments to liberty and equality. Certainly, the cybernetic welfare state is as powerful a force in America as elsewhere, but here it seems limited by dialectical strains, the driving force of which involves the potent yet conflicting appeals of values linking equality, dignity, and liberty. This dialectical strain constitutes the source for tension between two important postures, one of which we characterize as the triumph of the therapeutic, the other we interpret as an unfolding of American value codes that generate a counter emphasis on accountability and responsibility. We will return to this issue subsequently.

Considerable complexity in these developments follows from the blurring of lines dividing public and private domains, as well as from the blurring of institutional boundaries and the interpenetration of organizations and agencies. It is clear that institutional boundaries between government and private business, government and universities, and various government bureaus have not disappeared; but they have been much less sharply drawn than in the past. The role of the federal government at local and state levels is reasonably well known, if inadequately understood by Washington itself. Indeed, social scientific evaluation research has played a major role in giving the federal government influence in domains over which it cannot claim formal jurisdiction. For example, setting guidelines and evaluative standards in professional fields such as education has gone far toward pressing local and regional authorities in the direction of a national policy. Evaluation research and the idea of public accountability are also elements in the increasing interpenetration of institutions. The role of government in the universities obviously has increased dramatically, not only through the expansion of the state-sponsored sector as against the sector of private institutions, but also through governmental assumption of an active role in shaping higher educational policies through guidelines for personnel selection and through standards for federal grants and contracts. Even though universities continue to preserve autonomy to a certain extent, they also clearly act as agents of the government in relation to students and other clienteles.

It is not at all difficult to document the same point of institutional interpenetration in relation to the complex of government and private business. Large corporations have in certain instances developed such complete dependence on governmental appropriation, for example in the defense, aerospace, and related high technology fields, that they appear from the outside as quasi-extensions of government. It is also true that the political weight of such corporate employers, having critical economic significance for entire regions and acting as organizational reservoirs of scarce technical expertise, burdens the government with responsibility for their maintenance. For example, the financial difficulties of Lockheed corporation were ameliorated through decisively important assistance and contracts from the government.

The interpenetration of both public and private agencies has also created an understanding of complex organizations different from earlier models in which conceptions of monolithic, hierarchic bureaucracies prevailed. One person may well act now in the intersecting force fields of several organizations, linked to the pursuit of tasks through often complex and indirect methods of coordination. This kind of experience in post-modern complex organizations, where the pursuit of tasks has to be carried out within the contexts of great complexity and across often intersecting institutional boundaries, differs drastically from both the experience of individuality in the marketplace and the experience of the classic bureaucratic situation. There is a need for unprecedented flexibility here, which illustrates paradigmatically the complexities of relating individuals and collectivities in post-modern contexts.

Another aspect of this complexity involves the limitation of liability through governmental, collective absorption of risk. It can also be described as the diffusion of responsibility into anonymous systems. This tendency is mirrored in the public readiness to perceive possibly unrealistic limitations to potentially serious threats. For example, the rising scarcity and cost of energy may well have a drastic impact on American life styles, yet public attitudes seem buffered against the perception of risk. Many still have naive confidence that "they" will do something to absorb the risk or even make it disappear.

AN ENCOMPASSING CONTEXT

The value movements of our time, like those of other ages, often are limited by complex dialectics, not infrequently deflecting from a value in the process of its attainment. The value complex of equality-liberty-dignity, so prominent today in an arena of competing groups, is not exempt. This complex may well be of world-wide importance. For it is in relation to this complex that Raymond Aron pointed to the dialectic of equality and hierarchy as one of the central conditions of contemporary society. He notes that rationalism is the common source of both the egalitarian ideal and the boundless desire to dominate nature through technology. Yet "the will to produce as much and as efficiently as possible is, however, inherently alien to the concern for justice and to other humanitarian considerations" (Aron, 1969). He suggests that the production of wealth requires rational organization, efficiency, and that means hierarchy and authority. It militates against the desire for equality, even if the resulting hierarchy were strictly based on abilities. "Ideally this hierarchy would have to conform to a hierarchy of abilities. But, if there were a vast and evident gap between the least able and the most able (or skilled, or gifted), would the collective conscience accept such inequality?" (Aron, 1969).

It is in contradiction within the egalitarian ideal itself as well as in contradictions between egalitarianism and other collective priorities that Aron sees a central source for the instability of modern societies. "They are continually being unsettled by the progress of science and technology and by the dissatisfactions of men. Order in such societies is always provisional; it is not so much *in* process as it is *a* process" (Aron, 1969).

There are, indeed, the inherent limits to equality and hierarchy of which Aron speaks. Yet another tension that needs attention in relation to this value complex involves conflicting demands for identity and difference. Movements for minority rights as well as movements for ethnic and racial equality, inevitably strive for positive self-identity and for the preservation of positively valued differences, while simultaneously presenting claims for equal rights in dignity and life chances. The black movement in the United States is an excellent example of this point. The demand for equal rights does not mean denial of cultural and ethnic differences. Indeed, demands for equality in participation are accompanied by claims for dignity in the recognition of differences.

One complexity in this context is that the search for identity and emotional identification enhances group differences in certain respects while decreasing them in others. There has been a well-known revitalization of ethnic identities including governmental recognition and political rhetoric concerning their virtue. In some respects, then, claims for equality mean convergence of all groups on the prevailing standards and styles of the public domain of the larger society; in other respects, the same process means divergence in the maintenance and possibly even self-conscious cultivation and enhancement of ethnic cultural differences.

In a most complex fashion this process also is linked to the shifting relationships between public and private domains and between individual and collectivity. The formal, legal constitutional framework has been employed most explicitly in the struggle for minority rights in America. Through it, concrete definitions of civil liberties and individual rights have been articulated that have profoundly influenced cultural models for the construction of individual identities. This central value code articulated in historically specific contexts is of enormous and repeated significance in the American experience. And civil religion once again is playing a major, albeit indirect, role in the process. Not only is it apparent in the prominent role of religious leaders in the civil rights struggle, but it also is significant in making religious ideas of conscience and freedom an increasingly strategic line of demarcation between loyalty demands that the state may legitimately make of individuals and their right to be exempt from them or even to resist them. A peculiar adaptation of central American values about the relation between individual and collectivity appears to be emerging. Debates about the protection of privacy in increasingly prevalent contexts of publicly available information about individuals, the extension of citizenship models of individual rights into the structure of the family and marital relations, as well as the extension of entitlements to include the right to be educated and the like are all aspects of this historic process. Indeed, despite the similarities between American and European experiences, we feel that there are critical differences between Europe and the United States in this regard that will produce significant cultural differences in their adaptation to the post-modern future.

    These reflections bring us close to our next theme, the peculiar salience of the problem of identity, which involves some fundamental consequences of patterns of knowledge application.

THE ISSUE OF IDENTITY

    In the context of rapidly shifting power systems, not only within the United States but also around the world, and in the context of rapid cultural change with its attendant reorientation from fixed structures to processes, it no longer is possible to assume that personal and collective identities are stable and continuous. At the very least, stability cannot be taken for granted. There is a great deal of reflectivity about contemporary society, a very considerable emphasis on social measurement and, as a consequence, on the objectification of social experience. Of course, one finds maxims and principles for the guidance of the soul or the attainment of virtue or the attainment of mental health and happiness in all complex cultures, but it is only in contemporary American society that the idea of personal identity as something deliberately to be worked on or transformed has gained credence and popularity. Many Americans today believe that they ought to take themselves and their identities as objects capable of deliberate modification. There is little doubt that the explicit concern with identity has intensified in the recent era. One major source for the concern lies in the visible processes of identity construction and reconstruction, both at individual and collective levels, occurring in political and legal contexts. Thus, while the concern takes political and cultural forms grounded in past trends and value codes, it also has found a peculiar and special expression in post-modern American cultural movements.

Marx has argued elsewhere that post-modern cultural movements represent a departure from the pattern of mobilizations so prominent in the past (and continuing into the present) that aimed at institutional change, reform, or revolution. American cultural movements in the contemporary era provide an encompassing context within which new models for adult identity and social consciousness are constructed, legitimated, and internalized. The critical mechanism for the accomplishment of these often explicitly stated goals is the

> ...homogeneous primary group context containing intimate peer relations which provides a structural, legitimate moratorium on normal institutional role performances and responsibilities as well as a cultural free-space which encourages experimentation and innovation with prevailing symbolic codes.
>
> *(J. Marx, 1979)*

Under these conditions a high level of trust is built up so that an ideological reinterpretation of reality can occur that affects both the personal-private and the collective-public spheres. Intensive small group contexts for these processes, ideological primary groups, are probably the principal resocializing contexts for adult identity models and thus crucial mechanisms for the transformation of personal and collective consciousness.

> But even more importantly, these group experiences transform psychologically sophisticated "moderns" appropriate to industrialized institutional structure into sociologically sophisticated "post-moderns" suited to the rigidity of post-industrial institutional structures by virtue of becoming, in a non-technical way to be sure, aware of the socially constructed nature of reality.
>
> *(J. Marx, 1979)*

In this view, the individualism connected with the process of industrial modernization produced self-conscious, psychologically oriented people for whom a central source of stability and continuity derived from striving for a sense of unitary, continuous personal identity and self-hood. The very high mobility and rapid cultural change in current society may well have rendered this conception of psychological man somewhat less plausible. It appears that a more sociocultural conception, however diffuse, is gaining currency. This seems to be particularly apparent in those groups that self-consciously attempt to construct new personal identity models. Ideological primary group processes that lead to adult identity transformation in contemporary cultural movements do generate an awareness, however rudimentary, of the social construction of reality.

Ideological primary group processes and cultural movements make members aware that formerly imposing edifices of institutional role constraints are not necessarily givens that inhere in the nature of things. Rather, they come to be seen as socially constructed entities that can be altered. Indeed, in some instances, the belief has been cultivated that the mere withdrawal of commitment and the alteration of consciousness brings about significant structural as well as personal change. Many participants in contemporary cultural movements have thus come to appreciate the relativity of perspectives and viewpoints. The peculiar notion that to work on oneself for the purpose of growth as one of the main, if not the only, objectives worth striving for, has led to the "psychological utopianism" (Manuel, 1965) characteristic of contemporary American cultural movements, especially in the upper middle class.

In a sense, this phenomenon represents an attempt at knowledge application in its most extreme form. Much of the ideological content of recent cultural movements has been shaped by reliance on simplified, popularized behavioral and social science. We will return to the role of behavioral science in a different context, merely suggesting here that such self-absorbed concerns with identity quite possibly lead to a cultural dead end.

RISK ABSORPTION AND THE IMMANENCE OF SOCIETY

Our brief treatment of the major cultural concern with identity has shown, we think, that post-modern culture can become in a peculiar sense introverted, turned inward on itself. It is in this sense that we speak of the immanence of the post-modern pattern. This may contradict the fact that there is considerable awareness of the limits society encounters in its struggle with nature, an awareness of the borders between the human world and the vast natural ecology in which it is embedded. The debate about limited resources, the concern with pollution, the problem of energy, all relate to this issue and point beyond the boundaries of society itself. However, in spite of all this, contemporary American society seems peculiarly self-consciously concerned with itself and thus self-absorbed. We suggest that this phenomenon, so apparent in the often unquestioned trust in the efficacy of government and the economy as well as in the tendency to mistake the social world for the world itself, has its roots in the very nature of post-industrial society and in its peculiar pattern of uncertainty management through the spreading of risks.

Daniel Bell's (1973, 1976, and 1977) observations provide a basis for the first point. He distinguished between pre-industrial, industrial, and post-industrial societies. He viewed the pre-industrial world as primarily a "game against nature." In this pattern, the individual's experience of the world is conditioned by the vicissitudes of the seasons, the weather, the soil. The forces to be overcome in such a pre-industrial world are tangible, but they are also capricious, in many ways beyond human control.

In contrast, an industrial world represents a "game against fabricated nature." In an industrial world, people are tied to the machine, dwarfed by its size and power, yet also magnified by the enormous increases of energy that now are capable of being harnessed in the industrial process. In an industrial world, the forces with which one must deal are tangible as well as methodical and metrical. It is the world of enormous machines and machinery; a world in which one is no longer simply part of nature but, rather, controls, masters, and exploits those aspects of nature still experienced.

The post-industrial world emphasizes services: professional-client relations, bureaucrat-client relations, and the like. The post-industrial world remains an age of machine and technology, but one's relation to them becomes much different from what it was in previous eras. Instead of the obvious dominance of gigantic machinery, modern machinery, especially automated systems, appear embedded in the visible social contexts that they are designed to serve. Indeed, machine design reflects the point. Modern design often shows impressive adaptations to the needs and comforts of the human operator. One only needs to think of the precarious manner in which engineers perched on ancient steam locomotives to see the difference. It is most important that, as Bell's argument shows, the post-industrial world is in large measure a game between persons. In a sense, this type of society and economy is peculiarly social and turned on itself.

Bell describes the most important changes in post-industrial society as involving the decline in the relative significance of manufacturing and the increased significance of the service aspect of the economy. A century ago, it was possible for social observers to discern the outlines of the change from an agrarian to an industrial society. In our time the structure of post-industrial society is becoming visible.

> Today in the United States, 65 out of every 100 people are engaged in services; by 1980 about 70 out of every 100 will be employed in that sector. And these are new kinds of services, not those characteristic of an agrarian society (largely household servants) or an industrial economy (largely the auxiliary services of transportation and utilities, as well as some banking) but "human services" (the expansion in medical care, education, and social welfare), professional and technical services (research, planning, computer systems), and the like. This has meant that the fastest growing segment of the American labor force has been the professional, college-educated people. Today the professional technical and managerial occupations make up 26 percent of the labor force, clerical workers, 18 percent, semi-skilled workers, 17 percent, and skilled workers only 13.5 percent.
>
> *(Bell, 1975a:56–57)*

Bell continues by pointing out that even in industry itself there has been considerable change leading to the replacement of both unskilled and semi-skilled workers.

> The man on the production line is giving way to the man who watches the dials or the man who comes in, as a skilled worker, to repair the machine. Repairmen and foremen accounted for 75 percent of the growth in skilled jobs since 1940.
>
> *(Ibid.)*

The very structure of work in post-industrial economies emphasizes the web of social interrelations. This appears to us one source of the great concern of post-modern cultural systems for problems of personal and collective meaning. It is in this rather special sense that we speak of the peculiar immanence of contemporary culture.

The other source of this characteristic derives from the very success of certain strategies for the management of uncertainty. The effort to limit risk through encompassing and complex insurance programs, especially public ones, has provided more than just an economic floor below which the condition of the aged and poor should not be allowed to sink. Risk spreading as an uncertainty management technique does not, of course, remove individual uncertainty entirely, but lessens its threat by limiting the dangers of negative consequences. Quite naturally, this political strategy of risk spreading has produced an attitude of *entitlements*, the expectation of rewards and privileges as a matter of course. The existence of social security and insurance programs, the manner of actuarial thinking involving probabilistic assumptions about life outcomes, has in a real sense affected the structure of contemporary common sense. It is taken for granted, in an attitude of trust, that extreme hazards and catastrophes can safely be excluded from practical consideration. This aspect of uncertainty management, built into strategically important institutional structures, further supports the salience of the post-modern cultural domain.

THE APPLICATION OF
BEHAVIORAL SCIENCE

Given our observations about the prominence of the issue of identity and the contemporary preoccupation with the immanently social, one can readily see a rationale for the remarkable pervasiveness and scope of applications of the behavioral sciences. Indeed, the behavioral and social sciences have undoubtedly found a more receptive market in the United States than anywhere else in the world. Technical, professional applications are only a relatively small aspect of the matter. Applied behavioral and social sciences, certainly not free of distortions, have assumed an unprecedented role in what might be termed the "consciousness industry" (Enzensberger, 1974). The range of professions dealing with "people processing" (Wheeler, 1966) in one form or another is enormous. It spans the realm of medicine, most notably psychiatry and clinical psychology, and includes many forms of legal practice, psychiatrically oriented social work, counseling, and others. The domain of the human service professions includes corrections officers as well as schoolteachers, vocational counselors, and encounter group organizers. Certain sectors have taken on the characteristics of a profitable industry.

All of these occupational roles use applied behavioral and social scientific knowledge, often for the purpose of changing some aspects of the lives or identities of their clients. This complex of helping-healing-human service professions is not only well anchored in governmental structure but also is linked to powerful mass media of public communication. The emergence of this consciousness industry, in fact, required the technological developments that led to the present position of the electronic mass media. Yet it extends far beyond their confines. It includes popularizations of psychiatry, encounter groups, and organizations, personal advice columns in newspapers, as well as the faddish literature on personal growth, success, creative divorce, open marriage, and the like. An enormous number of magazines with mass circulation are relevant in this connection, including *Psychology Today*, *Transaction-Society*, *Behavior Today*, *Cosmopolitan*, *Ms.*, and of course *Playboy*. The authoritative figures in this domain range all the way from preachers and moral entrepreneurs to serious scholars.

A great deal thus is being written and said about what conduct ought to be like, how personal or interpersonal problems are to be defined and treated, and how personal identity and maturity are to be attained. Far beyond being purely technical and utilitarian, this use of the behavioral and social sciences relates closely to the construction of new cultural models and the reconstruction of old ones. Naturally, fads and fashions are prominent in a sector of culture production so dependent on circulation-oriented media. We have little doubt that the aggregate impact of these efforts is enormous, especially considering the profound needs for guidance and wisdom that are pervasively experienced throughout contemporary society. The shallowness of much of what is being produced, its routinization, and often garish marketing should not distract from the historical importance of the phenomenon. In spite of all the diversity, one can discern certain thematic patterns, such as the recent emphasis on growth, which often means mere change for change's sake. As is so often the case, a counter-trend of critique seems to be appearing, pointing to the trail of broken commitments and the emptiness of disillusionment following from the pursuit of change without a larger purpose. One major theme has tended to characterize most thinking in this mode; it is the theme of the therapeutic, to which we now turn.

THE TRIUMPH OF THE THERAPEUTIC

One dominant model for the application of knowledge in the interpersonal domain of the helping-healing-human service professions, and in related areas of policy, is the medical, therapeutic model. The very persuasiveness and fashionableness of this model has made it one of the primary modes of orientation in dealing with problems of broadly divergent natures. In this conception, the highly directive, authoritarian, and manipulative application of knowledge is made legitimate through the healing, curative intent. Responsibility is absorbed by the therapist; a feature of the model that makes it fit in with broader tendencies toward the diffusion of responsibility, spreading of risk, and the management of uncertainty.

The ascendency of this therapeutic conception has been predicted for some time by astute observers of Western civilization. Thus, Reiff records (1966:24):

> In a passage of *Italienische Reise* (Stuttgart, 1862) dated May 27, 1787, Goethe remarks: "Speaking for myself, I too believe that humanity will win in the long run: I am only afraid that at the same time the world will have turned into one huge hospital where everyone is everybody else's humane nurse."

Indeed, Reiff (Ibid.:24) himself believes that the therapeutic is becoming one of the normative institutions of the emerging culture.

> Where family and nation once stood, or church and party, there will be hospital and theater too, the normative institutions of the next culture. Trained to be incapable of sustaining sectarian

satisfactions, psychological man cannot be susceptible to sectarian control. Religious man was born to be saved; psychological man is born to be pleased. The difference was established long ago when "I believe," the cry of the ascetic, lost precedence to "one feels," the caveat of the therapeutic. *

Reiff goes even further in that he argues that there is a shift from an emphasis on control of emotions and impulses to an emphasis on release.

> Viewed traditionally, the continuing shift from a controlling to a releasing symbolic may appear as the dissolution of culture. Viewed sociologically, the dominance of releasing motifs, in which the releases themselves evolve as new modes of control, with patterns of consumption as cur popular discipline, implies a movement of Western culture away from its former configuration, toward one in which old ideological contents are preserved mainly for their therapeutic potential, as interesting deposits of past motifs of moralizing. *(Ibid.:24)*

The extension of the idea of therapy to the interpersonal domain, and indeed to the domain of political solutions for social problems, stems from the Freudian revolution in psychology and psychiatry. In the United States this movement was powerfully propagated by the mental health movement, with its vast extensions into American social life through institutions, groups, and associations. Indeed, one might even argue that the *therapeutization* of post-modern Americans has been part of the larger cultural transformation in which medicine and psychotherapy have limited the role of both law and politics, on one hand, and religion, on the other, as definers of morality.

Regardless of their therapeutic or other actual benefits, intensive group techniques and procedures have become a regular feature of the contemporary American scene. An extremely conservative estimate has argued that more than 60 million Americans have participated in some kind of sensitivity training, encounter, or "personal growth" group (Maliver, 1971). Moreover, intensive group techniques and procedures have been adopted by quite divergent contemporary movements, although they are most obvious and pervasive in the

*From Philip Reiff, *The Triumph of the Therapeutic,* Harper & Row, Publishers, Inc. Reprinted by permission.

Feminist Movement (Mitchell, 1971; Freeman, 1972 and 1973; Cherniss, 1972); the communal movement (Kanter, 1968; Jacobs, 1971; Bart, 1971; J. Marx and Seldin, 1973a and 1973b); and the Jesus Movement (Mauss and Peterson, 1973). These techniques and procedures were given a certain substantive content through the various branches of humanistic psychology that fused nondirective counseling with existential psychotherapeutic concerns in the late 1950s (J. Marx and Seldin, 1973a).

It is significant that these techniques represented the outcome of explicit attempts to apply behavioral science knowledge for the purpose of modifying interpersonal relations and personal conduct within a framework that combined therapeutic and educational goals. Intensive groups are so popular because they serve as replacements for the positive communities that Reiff (1966) described as necessary anchorage points for the construction of identity, replacements that appear convincing in the fluid social and cultural context of the present. Of course, there are limits inherent in the group-centered approach to instant intimacy. Where the attitude of the group includes a refusal to make ultimate commitments or final choices, so that the search for authenticity becomes forever elusive, frustration and disenchantment necessarily result. Indeed, some of the recent movements of religious resurgence may benefit from this circumstance.

Further, the very ascendency of therapeutic models of thought calls for a dialectic counter-trend. Not only does the transfer of responsibility (in this model) to the superior knower (therapist) conflict with central themes in the American value core, but also the application of this pattern of thought creates perceived injustices and movements for redress. One example has occurred in the criminal justice system in which the reliance on indeterminate sentences in conjunction with therapeutic emphases on rehabilitation has created manifest inequities in how offenders are dealt with and has given inordinate powers to professionals ill equipped to justify these powers. The current movement for reform quite directly attacks therapeutic conceptions and replaces them with concerns for equity in the definition of punishment. Similarly, the mobilization of groups for political action that were targets of therapeutically conceived policies for the solution of social problems has dramatized the questionable nature of the approach. The therapeutic model of knowledge application, when extended in the manner in which it has been applied, can be a cloak for manipulative intervention.

ACCOUNTABILITY, AUTONOMY, AND THE THEME OF THE RESPONSIBLE CITIZEN

Quite explicit attacks on the utilization of therapy as a pervasive mode for dealing with human problems have dramatized the theme of accountability and the value of autonomous decision by well-informed citizens. Possibly most dramatic is the introduction of the doctrine of *informed consent* into the realm of psychiatry itself. The recipient of treatment, in this conception, must make responsible judgments and consent autonomously to the help professionally offered. Risks thus are to be shared between professional and client—with significant legal consequences. More important in our view is the explicit reassertion of certain fundamental American value themes clustering around the notion of the rational, responsible citizen. We already have pointed to the convergence of various normative concerns involving individual privacy, consumer protection, civil rights and liberties. In the ideal form, this pattern holds out a rather different model for the application of knowledge, one that sees the distribution of knowledge as a prerequisite of the individual's capacity to make informed and responsible judgments. In spite of the theme of limiting liability and risk, which we have described as one general trend, there is here an insistence on individual responsibility.

Any such normative conception works with fictions. Exactly where the capacity to bear responsibility realistically begins and ends is hard to say. Yet the reassertion of what we have described as the basic American value code is hard to overlook. It has very considerable consequences for the manner in which knowledge can be applied, not only in professional-client relations, but in the policy arena itself. The demand for transparency in the formulation of policy, exemplified by "sunshine" laws, the insistence on accountability of agencies performing public duties, all are aspects of this trend.

Indeed, the very structure and autonomy of epistemic communities may be undergoing a slow process of change in relation to these value movements and the inherent transformation resulting from increasing reliance on explicit, codified, and therefore replicable knowledge.

As knowledge production, distribution, and use have become such visible and central issues in contemporary economic and political life, it is inevitable that demands for their governance will arise. Indeed, the theme of accountability contains such demands, posing formidable problems. The American knowledge system is not centralized and monolithic, in spite of its increasing reliance on the federal government. Concerns for knowledge production and use tend to be tied to substantive domains of policy, vying with each other for priority and attention. The institutional structure includes relatively autonomous universities and professions, and relates to a multiplicity of decision centers. The creation of uniformity in science policy or, even broader, in the governance of the knowledge system, appears to us impossible in the American context. Yet the pluralistic solution is merely a framework for an unceasing contest of groups. It may well be that the primacy of the political will assert itself over any ideas of technocracy, however concealed.

CONCLUSION

In this concluding chapter, we have sketched some ways in which the matters of our immediate concern, the knowledge system and especially knowledge application, are embedded in and related to larger social, political, economic, and cultural contexts. In doing this in a speculative vein, it has become clear that the issues of knowledge use cannot be conceived of as limited to a narrow technical or utilitarian sector of social life and social thought. Indeed, we have emphasized connections to subtle cultural and psychological processes, like those encountered in the forming and changing of conceptions of identity and cultural movements. It has become increasingly clear how difficult the task of the analyst is in separating an object of study from the broad, confusing, and vital context of social life.

We must conclude with one final point about the significance of the knowledge system: The endeavor of science and scholarship always points beyond any given social form and cultural configuration. The growth of knowledge is, in a sense, society's frontier with nature and the future. We meant to focus on the application of knowledge and yet we could not avoid writing about basic science and scholarship as well. There is, to be sure, a powerful drift toward a ruthlessly utilitarian misconception about what a knowledge system in its basis really is. Restrictions imposed by rigidly utilitarian, efficiency-minded concerns, no matter how legitimated in the name of accountability and the like, may well erode the very foundation of the knowledge system. All knowledge application rests ultimately on the vitality of science and scholarship, but beyond this there is inherent in the knowledge system a capacity for the rigorous, rational critique of society and culture. It is of the essence that its vigor be preserved and that cultural free-spaces in which it can flourish be protected.

PART
IV
Bibliographic Notes

It seems somehow presumptuous to suggest a set of references in these bibliographic notes to Part IV, as it consists of but a single chapter entitled "Speculations on Sociocultural Change." Nevertheless, a number of works strike us as having particular significance for this subject. The work that unquestionably has the most revolutionary implications for a sociological understanding of morals and cultures is Philip Reiff's prophetic and penetrating *The Triumph of the Therapeutic* (1966). Similarly, Daniel Bell's article on "Teletext and Technology" (1977) represents the most imaginative and comprehensive set of suggestions as to the manner in which emerging technology and bodies of specialized knowledge and information can be expected to affect the post-industrial social structure and post-modern culture. Finally, the interested reader is referred to the Postscript to our Appendix for some additional references accompanied by a brief discussion of recent works dealing with social and cultural changes in the modern professions.

Appendix

On the Concept *Profession*

Perhaps the sole issue on which the sociological literature dealing with work groups is unambiguous in that there are certain occupations that can be considered professions, although what occupations fit this category and what specific characteristics warrant using this label are the subject of considerable controversy. It has been common among those who have attempted to resolve the issue of "what is a profession" to formulate a list of qualities or attributes that seem to characterize all those occupations considered professional. As Table I demonstrates, although a group of attributes has emerged that fairly regularly appears in the definitions of various authors, the listings are seldom if ever, identical. Even more disturbing is the fact that the various authors provide conceptually rather impoverished justifications for the particular attributes that are presented. The methodological strategy underlying such approaches is that by focusing on the obvious, established professions, such as medicine and law, one can isolate their distinguishing characteristics and/or developmental pattern, and thereby construct a model of what professionalism must be. The kinds of attributes that result from this dubious methodology are revealed in Table 1. Among the attributes most often discussed are: an abstract, theoretical knowledge base; an esoteric, specialized, technical skill; a long training and socialization; strong professional subculture and ideology; lifelong commitment to a structured career; autonomy of action; a formal

Table I. Sociological Definitions of Profession by Distinguishing Attributes (Phrase Used by Author Paralleling Attribute)

	ATTRIBUTES			
Author[1]	Theory	Skill	Sanction	Codes
1. Durkheim (1902)				
2. MacIver (1922)	Cultural interest of profession	Technical interest of profession	Extrinsic interest of profession	Professional codes
3. Carr-Saunders (1928)	Intellectual study	Skilled service advice	Practitioner qualification, upward status	Standards rules
4. Timasheff (1940)			Decentralized public service	
5. Hall		Specialties	Prestige	Code of ethics
6. Mills (1951)		Exercise professional skill	Money in fact and symbol	
7. Parsons[2] (1952)		Valued cluster of roles	Independent trusteeship	
8. Foote (1953)	Body of knowledge	Specialized technique	Status by community	
9. Caplow (1954)			Political agitation for public support	Code of ethics
10. Greenwood (1957)	Systematic theory	Authority	Community sanction	Ethical codes

	ATTRIBUTES			
Author[1]	*Culture*	*Service*	*Boundaries*	*Career*
1. Durkheim (1902)	National occupational corp.			
2. MacIver (1922)	Integration	Ideal of service	Demarcation	
3. Carr-Saunders (1928)	Dominant professional association	Fee or salary		Training
4. Timasheff (1940)				
5. Hall	Formal association		Own professional characteristics	Own training
6. Mills (1951)	Guide-like system to control			
7. Parsons[2] (1952)				Full-time
8. Foote (1953)	Association of colleagues			Career
9. Caplow (1954)	Professional association, definite membership		Change name	Controlled training, selection
10. Greenwood (1957)	Culture	Career concept, "calling"		

Table I. Sociological Definitions of Profession by Distinguishing Attributes
(Phrase Used by Author Paralleling Attribute)

	ATTRIBUTES			
Author[1]	*Theory*	*Skill*	*Sanction*	*Codes*
11. Goode (1957) (1960)	Abstract knowledge			
12. Hughes (1958)	Research	Esoteric service	License, mandate	
13. Friedson (1960)				
14. Blau & Scott (1962)	Neutral	Expertise, achievement		Universal
15. Barber (1963)	Generalized, systematic knowledge		Prestige symbol of achievement	Internalized codes
16. Kornhauser (1963)	Intellectual content	Specialized competence	Autonomy	Standards of excellence
17. Ben-David (1964)	Intelligentsia		Ensured occupational advantages	Codes of behavior
18. Strauss et al. (1964)		Expertise	Legitimized claims	Set goals, standards
19. Wilensky (1964)	Systematic knowledge or doctrine	Technical competence	Mandate for monopoly	Professional norms

1, 2 (See p. 347 for complete footnotes to Table I.)

	ATTRIBUTES			
Author[1]	*Culture*	*Service*	*Boundaries*	*Career*
11. Goode (1957) (1960)	Professional community	Service orientation		Prolonged training
12. Hughes (1958)	Culture, lasting social organ			Socialization, enduring membership
13. Friedson (1960)	Control by colleagues, patients			
14. Blau & Scott (1962)	Colleague control	Client interest		
15. Barber (1963)		Self-control	Community, not self-interest	
16. Kornhauser (1963)	Institution, corporate body	Responsibility		Commitment
17. Ben-David (1964)	Vocational subculture, esprit-de-corps			
18. Strauss et al. (1964)	Occupational groups	Areas of human concern	Encroachment conflict	
19. Wilensky (1964)	Professional association	Service ideal		Long proscribed training

occupational association; control over training and legal licensing; and a code of ethics and/or a client-centered orientation—the service ideal.

Related to this conceptual/definitional morass surrounding the concept of profession is the recurrent, but increasingly popular, generalization that the entire occupational structure, the labor force as a whole, is becoming professionalized. (See Foote, 1953, and Vollmer and Mills, 1966, for the clearest arguments for the professionalization of many emerging occupational groups.) This label has been loosely applied to increasing specialization and transferability of skill, the proliferation of objective standards of work, the spread of tenure arrangements and careers regulated and supported by a colleague group, licensing or certification, and codes of ethics for service occupations. Table II presents the arguments for and against the contention that there has been a professionalization of labor, taken from Wilensky's brilliant critique of "The Professionalization of Everyone?" (1964).

The arguments for the view that labor has become professionalized rest on four attributes that Foote (1953) feels characterize professions: (1) a theoretical knowledge base for specialized techniques, (2) transferable skills, (3) careers regulated by a colleague group, and (4) a code of ethics. Using Wilensky's (1964:139) arguments against the position that labor has become professionalized, it is easy to demonstrate the frailty of Foote's four criteria, and their inability to discriminate what both the lay public and professional groups themselves consider professions from other occupational groups. First, the skilled trades, astrology, embalming, and many other nonprofessional occupations involve a well-developed knowledge base for their specialized techniques. Second, transferability is often absent in the established professions. As Wilensky (Ibid.) notes: "the house council and other staff experts have skills which are bound to a particular organization; knowledge concerns the traditions, personalities and procedures unique to that organization" and are thus nontransferable. In contrast, many of the skilled trades and the crafts involve readily transferable skills. Third, electricians and plumbers possess careers to the same extent that physicians do. Unions have guaranteed that such workers will spend their entire working lifetime within the same occupation, and that they will move in an orderly manner through the various statuses and positions

Table II. *IS THERE A PROFESSIONALIZATION OF LABOR?*[a]

Labor Becoming Professional*	Labor Not Becoming Professional
1. More manual jobs involve a *specialized technique* supported by a body of *theory*—mathematical, physical, chemical, even physiological and social-psychological. Systematic on-the-job training leads to *upgrading* of machine tenders. "Almost every employee in the plants of Detroit will be an engineer of one kind or another."	1. Specialization is no basis for professional authority. High degrees of specialization prevail at every skill level, both in jobs which can be done by almost anyone (assembler in auto plant) and in jobs done only by the trained (surgeon). The link between manual work and theory is tenuous. While ultimately all labor is rooted in theory, the civil engineer who designs a bridge may know some laws of physics, the workers who build the bridges do not. If by upgrading we merely mean ability to learn by virtue of previous experience and challenge, then IQ becomes the criterion for professional.
2. *More manual jobs involve transferable skills.*	2. *Transferability of skills* is limited in most manual jobs outside of the traditional crafts; much of the new technology requires training for the specific system or machine in the particular workplace. Such transferability is often absent in the established professions anyway. E.g., the house counsel and other staff experts have skills which are bound to a particular organization; their knowledge concerns the traditions, personalities, and procedures unique to that organization.
3. More manual jobs now provide *careers* regulated and supported by a colleague group. Unions try to	3. No evidence that colleague control of manual jobs is increasing. The pace of technological change is fast;

"reconstruct industry so as to assure every man a career." The evidence is in union demands for:

a) *Seniority in promotions*—to assume that everyone will wait his turn and so eventually reach "a job which matches his highest powers—one object of a career." Like the system of rank and pay of the professor.
b) *Demand for continuous income*—wage and work guaranties, pension and welfare programs, salaried status.

the changes are administered mainly by employers. This makes it unlikely that manual jobs, comprising a declining fraction of the labor force, will provide stable "careers" in any sense of the term.#
Most of these ways to increase the job security of high-seniority workers are traditional; they have not prevented the wiping out of obsolete crafts or unstable employment among low-seniority men. If salaried status were a criterion of professionalism, we would have to call the fee-taking doctor non-professional and the office clerk professional. If stable attachment to the enterprise is the criterion, this goes with a decline in colleague control.

4. *An objective and fair set of rules and standards* enforced by grievance procedures and arbitration under union contract is the *equivalent of professional codes of ethics* supported by public trust. Moreover, unions show an increasing responsibility toward the consumer—a new concern with the impact of wages on prices, laws to protect consumers, etc.

4. Negotiated plant rules are not focused mainly on quality of product or of work performance; they are overwhelmingly and properly concerned with protection of employees rather than the public. The public-relations programs of both unions and management should not be mistaken for their hard-core policies in contracts and daily life.

*The arguments are adapted from Nelson Foote, *op. cit.*, who states them in their sharpest form. Quotations are from Foote.

#H. L. Wilensky, "Orderly Careers and Social Participation: The Impact of Work History on Social Integration in the Middle Mass," *American Sociological Review*, XXVI (August, 1961), 521–39.

a From Harold L. Wilensky, "The Professionalization of Everyone?" *The American Journal of Sociology*, Vol. LXX, Number 2, September, 1964, p. 139.
Reprinted by permission of The Chicago University Press.

within that career structure. Fourth, and finally, codes of ethics are affirmed by dry cleaners, truck drivers, and army recruits. It is a moot point whether the ethical codes of the physician, lawyer, and clergyman are more powerful influences on occupational behavior than are the codes of these less elevated groups. In short, even taking all of Foote's attributes together as a single set of criteria, there are still numerous nonprofessional occupations that fall into his "attribute net." The skilled trades are a clear case in point; plumbers, electricians, and many other craft occupations involve a knowledge base, transferable skills, careers regulated and supported by a colleague group (through the craft union), and a code of ethics.

We cannot hope to resolve the ongoing debate as to the definition of profession, or whether the term should be dropped altogether in favor of the concept of professionalization, which implies a continuum along which all occupational groups are assumed to move as they develop more and more of the characteristics of medicine and law. But in order to proceed with our argument concerning the role of professions in the organization, distribution, and storage of knowledge, it is necessary to circumscribe the boundaries of our subject matter for, surely, we do not mean to describe all occupational groups. For our purposes, Freidson's conception is a useful point of departure (1970:xvii):

> It is useful to think of a profession as an occupation which has assumed a dominant position in a division of labor, so that it gains control over the determination of the substance of its own work. Unlike most occupations, it is autonomous or self-directing. . . . In developing its own "professional" approach, the profession changes the definition and shape of problems as experienced and interpreted by the layman. The layman's problem is re-created as it is managed—a new social reality is created by the profession. It is the autonomous position of the profession in society which permits it to recreate the layman's world.

To this formulation, we propose to add some of the specification that Wilensky suggests in the following formulation (1964:138):

> Any occupation wishing to exercise professional authority must find a technical basis for it, assert an exclusive jurisdiction, link both skill and jurisdiction to standards of training, and convince the public that its services are uniquely trustworthy.

In this view, the marks of a profession are a successful claim to exclusive technical competence and the exercise of autonomous expertise, as well as adherence to the service ideal and its supporting norms of professional conduct. To assert that the job of the professional is technical means that it is based on systematic, theoretical, esoteric, and abstract knowledge or doctrine acquired only through long, prescribed training and socialization.

It is important to note here that technical does not imply scientific. That is, nonscientific systems of thought or knowledge can serve as a technical base for a successful or established profession, insofar as rigorous standards of training are enforced and required for certification, and the theoretical doctrines are well codified and systematized. An outstanding example of a case where nonscientific, yet technical, knowledge and doctrine serve as a basis for practice in an exclusive occupational jurisdiction is the ministry. Occupational mandates derived from morality or religion and sanctioned by public opinion or by the supernatural (as in the case of the legal profession and the priesthood, respectively) may achieve just as secure an exclusive occupational jurisdiction as mandates derived from science and sanctioned by law (as in the case of modern medicine).

In terms of the conception of professions we are advancing, the key features of these occupational groups are their autonomy and authority over an exclusive jurisdiction that, in turn, rests on some technical knowledge or doctrine. Thus, it is the exclusive possession of some abstract, systematic, esoteric knowledge or theory that is the basis for an occupation's claim to the technical competence and expertise that justify professional status and prestige associated with that status. But this is only one sense in which professions organize, distribute, and store knowledge. Even more important is how both the demands of professional practice as well as threats to exclusive jurisdiction and competition with other occupational groups structure the nature of the knowledge base of highly professionalized occupations.

The technical knowledge base of a profession must be sufficiently abstract, esoteric, and difficult to acquire to clearly differentiate it from common sense. But just as a knowledge base consisting of a vocabulary that sounds familiar and obvious to everyone is a weak basis for professional claims to exclusive jurisdiction based on *technical* expertise, so is a knowledge base so narrow that it can be learned as a set of rules by most people. When technical knowledge becomes so restricted and precise that occupational skills can be broken down into their component elements, and the sequences of tasks that make up a performance can be prescribed in minute detail, the job comes to require little understanding and/or judgment on the part of the worker. Such a job can be learned in a very short time, without extensive intellectual preparation; it rests on a set of easily routinized and disaggregated procedures. In short, there seems to be an optimal base for the technical knowledge that can justify professional claims to exclusive jurisdiction and autonomous expertise: it must be neither too broad, vague, and familiar nor too narrow, precise, and restricted. Social work, social science, education, and administration are occupations that have encountered difficulty justifying their claims to professional status, prestige, and rewards because they rest on a technical knowledge base that is too broad, familiar, and close to common sense. Electricians, plumbers, and many skilled workers have had difficulty in pressing their claims for the kinds of treatment accorded professionals because the technical knowledge on which the performance of their work rests is too restricted, specialized, and narrow. Thus, the growth and prestige of the established professions and the possible professionalization of large sectors of the labor force (however superficial) affects the scope, breadth, and organization of the kinds of technical knowledge that are sought in the (increasingly professionalized) post-modern occupational structure.

The technical knowledge or theory that is the basis for professional practice has a number of other characteristics that have implications for post-modern knowledge in general. Specifically, the technical knowledge that is the basis of professional work is to an important extent tacit; it rests on understanding complex relations and processes that cannot be fully reported. It is this element of tacit knowledge—acquired through experience and participation, rather than learned from books—that gives established professions their aura of mystery and occasionally gives the most pedestrian professional practitioner a charismatic ambience. The client public sees unfathomable mysteries in the tasks to be performed and the problems to be dealt with; mysteries that it is not given to the ordinary person to acquire. In fact, very ancient usages of the noun *mystery* included occupation, craft, skill, calling, etcetera. But the fact that the technical knowledge base of a profession is partly tacit has even more important consequences. That is, tacit knowledge is relatively inaccessible to direct critical examination and debate. Therefore, the tacit role-embedded professional knowledge base is also tenaciously resistant to revision and/or change. This partly accounts for the traditionalism and conservatism of the established professions like medicine and law. Thus, the "laying on of hands" in medicine is surrounded by an aura of mystery and charisma, and the legal emphasis on the "majesty and mystery of the law" is cultivated as a way of emphasizing the extraordinary qualities of lawyers. Both of these emphases refer in part to the fact that important components of the profession's technical knowledge are tacit. Wilsensky (1964:149) even suggests that professional expertise may be defined as such thorough knowledge about a subject that it is possible to communicate only a small part of it.

The argument so far has suggested that established professions play a critical role in organizing, distributing, and storing knowledge and that many post-modern occupations are becoming increasingly professionalized, leading them to acquire or construct a body of technical knowledge, theory, or doctrine that will justify their claims to professional autonomy, authority over an exclusive jurisdiction, and prestige as well as rewards. We have suggested that the optimal base of technical knowledge for a profession combines abstract-theoretical and concrete-practical understandings. Some of these understandings are explicit, such as the generalizations and classifications acquired from books and lectures; some are tacit and implicit, acquired through practice and experience. The explicit, abstract-theoretical component of professional knowledge, as well as the tacit-implicit elements in both theory and practice, require long training and experience. The long, arduous training, the tacit understanding of esoteric, technical matters, and the issues that are dealt with in professional practice (i.e., life and death, justice, truth and morality, the sacred and the profane) all combine to persuade laymen of the mystery of professional activities and lend an aura of charisma to the practitioner. Finally, an occupation based on technical knowledge that is either too general, vague, and familiar or too narrow, specific, and specialized is unlikely to achieve the exclusive jurisdiction and autonomy necessary to the exercise of professional authority. Only a technical body of knowledge that has close to optimal scope and breadth provides the inaccessibility to critical examination and investigation by laymen, and therefore provides the resistance to modification or change that is associated with the conservatism, traditionalism, and self-confidence of professionals.

These considerations point to one of the emerging paradoxes surrounding the technical knowledge base of professional practice and post-modern common sense. Rising educational levels among the general population have generated, we believe, a new (post-modern) kind of common sense, one that emphasizes understanding processes and procedures rather than substantive rules and information, and that stresses the importance of knowing how to gain access to the appropriate (professional) experts/specialists for dealing with a particular problem. Thus, post-modern common sense produces a greater willingness to utilize specialized professional services and includes knowledge of how to find and benefit from the service most appropriate to a problem. At the same time, however, post-modern common sense is accompanied by greater sophistication about professional matters, greater skepticism about the certainty and even the efficacy of professional practices, increasing cynicism about professional motivations and about the rationale underlying professional practices and traditions, as well as some actual exposure and sharing of the technical knowledge employed by the professional. In short, much of mystery surrounding professional activity is dispelled and the public becomes increasingly disenchanted. The result is a post-modern public that is more critical and less deferential, docile, and awed by the outcomes of professional intervention, yet more willing to utilize specialized professional services for any problem whatsoever. In fact, this trend has already reached the point where the public (at least certain segments of it, especially youth) seems to attribute more mystery and magical properties to certain nonprofessional (quasi-artistic) craft occupations—such as carpenters, potters, jewelers, etcetera—than to the highly rationalized, specialized, and bureaucratized established professions, which are viewed as organized groups of commercial vendors.

Professional services always involve the attempted solution of the concrete problems of individuals and this circumstance affects the kinds of technical knowledge organized, distributed, and stored by professional communities. Professionals differ from scientists in their reliance on applied rather than formal theoretical knowledge.

Thus, practicing professionals tend to use general principles to deal with concrete problems whereas scientists investigate concrete problems in order to test, elaborate, or arrive at general principles. Since professionals must focus on the practical solution of concrete problems, they are characteristically more concerned with applying knowledge than with creating or contributing to it. Indeed, the need to do something and intervene, irrespective of the existence of reliable knowledge, leads to something like disdain for highly abstract and formalized bodies of theory or doctrine. This, in turn, generates an emphasis on personal experience with individual cases, rather than a concern for general laws or rules of procedure. And this means that professions always rely on pragmatic reality tests and criteria for evaluating procedures. This commitment to action, to pragmatic assessments of the apparent results of concrete measures for dealing with individual clients, leads to an acceptance of the idea of indeterminacy of uncertainty, rather than the idea of lawful regularity. This emphasis on uncertainty also serves to justify the professional's pragmatic emphasis on firsthand personal experience, obtained with concrete individual cases. In short, it seems fair to suggest that the key feature of technical professional knowledge is a rather thoroughgoing ontological and epistemological individualism. This means that technical professional knowledge is likely to be unusually self-validating and self-conforming, because the assumption of indeterminacy minimizes the role of scientific knowledge as well as the role of other's opinions.

The fact that professional services are directed toward the solution of concrete individual problems has several consequences. First, the professional models developed to organize the technical knowledge guiding actual practices and techniques focus on the isolated case as the critical unit of analysis. Thus, in medicine the focus is on the disease (the diseased part or the disease entity) presented by the patient, rather than on the patient him/herself. (It is this narrow case emphasis engendered by the disease model that has recently brought extensive criticism on contemporary American medicine and resulted in the development of a new specialty, Family Practice, designed to rectify the overly restricted focus on treating the disease, rather than the patient.) Similarly, the legal profession is concerned with particular concrete cases, rather than an individual's general moral status. And even the entrepreneurial elite attending the prestigous Harvard Business School receive their managerial and administrative training via the case method. Such an emphasis on isolated cases is inimical to the kind of probabilistic reasoning and development of general concepts and principles so necessary for scientific advance. Second, the technical knowledge base of the professions encourages the organization of knowledge around problem areas, rather than in terms of logical premises, underlying assumptions, or epistemological *desiderate.* This means that the development of knowledge tends to proceed along the lines established by public and professional definitions of problems, and knowledge becomes valued only insofar as it can be shown to bear directly on the amelioration or resolution of these problems. Knowledge becomes almost exclusively instrumental.

FOOTNOTES TO TABLE I

1. Emile Durkheim, *Division of Labor in Society*, trans. George Simpson (New York: Free Press of Glencoe Inc., 1933), p. 26 of the preface; Robert MacIver, "Social Significance of Professional Ethics," reprinted in Howard M. Vollmer and Donald L. Mills, *Professionalization* (Englewood Cliffs, N.J.: Prentice Hall, 1966), pp. 51–54; A. M. Carr-Saunders, "Professions: Their Organization and Place in Society," reprinted *ibid.*, pp. 3–7; N. S. Timasheff, "Business and the Professions in Liberal, Fascist, and Communist Society," reprinted *ibid.*, p. 59; Oswald Hall, "Sociological Research in the Field of Medicine: Progress and Prospects," *ASR*, XVI (October, 1951), p. 643; C. Wright Mills, *White Collar* (New York: Oxford University Press, 1951), p. 136; Talcott Parsons, "A Sociologist Looks at the Legal Profession," *Essays in Sociological Theory* (Glencoe, Ill.: Free Press, Inc., 1954), p. 372; Nelson N. Foote, "Professionalization of Labor in Detroit," *American Journal of Sociology*, LVII (January, 1953), p. 372; Theodore Caplow, *Sociology of Work* (Minneapolis: University of Minnesota Press, 1954), pp. 139–40; Ernest Greenwood, "Attributes of a Profession," reprinted in Vollmer and Mills, *op. cit.*, pp. 10, 17; William J. Goode, "Community within a Community: The Professions," *American Sociological Review*, XXII (April, 1957), p. 194, and "Encroachment, Charlatanism, and the Emerging Profession; Psychology, Sociology, and Medicine," *American Sociological Review*, XXV (December, 1960), p. 903; Everett C. Hughes, *Men and Their Work* (Glencoe, Ill.: Free Press, Inc., 1958), pp. 78, 117–19, 140–42, 159; Eliot Freidson, "Client Control and Medical Practice," *American Journal of Sociology*, LXV (January, 1960), p. 382; Peter M. Blau and W. Richard Scott,

Formal Organizations: A Comparative Approach (San Francisco: Chandler Publishing Co., 1962), pp. 60–62; Wm. Kornhauser, *Scientists in Industry* (Berkeley, Calif.: University of California Press, 1962), pp. 1, 8, 11; Bernard Barber, "Some Problems in the Sociology of the Professions," Daedalus XCII (Fall, 1963), pp. 672–73.

2. Parsons' essay on the legal profession was selected, rather than the classic essay "Professions and Social Structure," for the defining attributes of profession because: (1) being the most recent, it appears more accumulative of the theorist's thought and (2) in it he explicitly attempts to distinguish professions. In the earlier effort, his purpose was to show the similarity, not to distinguish, between professions and business: they have similar characteristics (rationality, functional, specificism, universalism) and each pursues structurally defined goals of success (via altruistic expertise in one case and via acquisitive efficiency in the other). His characteristics distinguish occupational *gesellschaft* from *gemeinschaft*, not necessarily professions from other occupational groups. In his later essay, he mentions the importance of "independent trusteeship" on his way to examine the function of professions as "mechanisms of social control" (via permissiveness, support, noninvolvement, manipulation). It is this "independent trusteeship" distinction that can resolve differences between Parsons and Hughes, as reported by Peter M. Blau and W. Richard Scott, *Formal Organizations: A Comparative Approach* (San Francisco: Chandler Publishing Co., 1962), p. 63.

Postscript

The post-World War II American sociological literature on the professions has been dominated by one central theme: The extent to which it is possible to maintain that the entire society is becoming professionalized. The 1950s saw a veritable barrage of papers and books—exemplified by Nelson Foote's (1953) paper—asserting that labor in particular, if not all occupations and the whole society, were becoming professionalized. Harold Wilensky's (1964) paper on "The Professionalization of Everyone?" asserted a different thesis. Wilensky argued that the traditional professions were unique in that they were based on esoteric knowledge acquired through long, intensive training and possessed a service ideal as the core of their normative structure. In his analysis, the established professions differ from other occupations in that they have undergone a process of professionalization involving the institutionalization of training within universities, the formation of professional associations, monopolistic power over licensing, and the adoption of explicit ethical codes. These characteristics were seen as the critical prerequisites for the extraordinary degree of autonomy that is associated with the established professional occupations. The central meaning of this autonomy involves the authority and freedom for self-regulation and action within defined spheres of competence.

A little more than ten years later, Marie R. Haug (1975) published a paper entitled "The Deprofessionalization of Everyone?"

that questions the widely disseminated image of a professionalized society in which knowledge, especially specialized, technical professional knowledge and expertise, will be the dominant source of power. Her analysis emphasizes the erosion of the professional monopoly of specialized knowledge as a result of rising levels of public education and sophistication, as well as specific programs and movements for client education. In her argument, the general weakening of the professional knowledge monopoly, as well as the decline of public belief in professional good will, cannot help but undermine professional autonomy—the generalized authority and freedom for self-regulation possessed by professions—and hence, ultimately, the status position and deference accorded to the established professions. Indeed, Haug's thesis suggests that *both* technological and ideological trends are rendering the concept, status, and autonomy of professions obsolete. Her argument suggests that as traditional professions lose monopolistic control over their knowledge domain, their autonomy and authority become challenged, and demands for accountability and client rights emerge. The central consequence of all this is an erosion of the high prestige formerly enjoyed by the established professions. The outcome that Haug foresees is a "deprofessionalized" future in which professional expertise is stripped of superfluous claims to mystery, authority, and deference, and in which clients become knowledgeable consumers who question, compare, and skeptically evaluate all kinds of professional services. In contrast to Wilensky's argument that focused on the unique features and characteristics of the established professions and professionals, Haug's analysis focuses on changes and emerging developments in the stance of contemporary receivers of service (clients) vis à vis the professional providers of service.

Haug's analysis represents an important challenge to the notion of a professionalizing, post-industrial society suggested by Daniel Bell (1973 and 1976). As we have indicated elsewhere in this volume, such a conception views the emerging society as one in which the professional and technical class will be pre-eminent, with theoretical knowledge a central source of power. The scientific knowledge explosion and the exponential growth and technology produce major shifts in the distribution of the labor force, highlighted by the growth of the professional-technical class that controls scientific and theoretical information. Indeed, it is the command or monopolized control of scientific knowledge in the hands of a professional elite that leads to the new distribution of autonomy, privilege, power, and prestige in the projected future. In challenging Bell's conception, which is dominated by the professional occupations, Haug's hypothesis of deprofessionalization does not imply the end of specialized, technical knowledge and expertise. What it does suggest, however, is that members of the professional-technical elite will no longer be able to rely on traditional claims of mystery, authority, and deference to secure disproportionate amounts of autonomy, privilege, power, and prestige. More succinctly, specialized, technical knowledge and (professional) expertise will no longer generate the patterns of obeisance, trust, and awe that emerged at the end of the nineteenth century.

An entirely different perspective—and one that is extremely compatible and sympathetic to the viewpoint developed in this volume—has begun to emerge in a number of recent social historical studies of professionalization. These studies indicate that professionalism did not emerge, in the nineteenth and early twentieth centuries, in response to clearly defined social needs. Instead, the new professions themselves invented many of the needs they claimed to satisfy. They played on public fears of disorder and disease, adopted a deliberately mystifying jargon, ridiculed popular traditions of self-help as backward and unscientific, misleadingly legitimated themselves in terms of the mantle of science, and generally created or intensified demands for their own services. The most important case for the utility of the helping-healing-human service professions rested, not on their technical or scientific superiority and efficacy, but on their ability to control clients. This perspective emerges with particular clarity in Burton J. Bledstein's analysis *The Culture of Professionalism* (1976) and in Magali S. Larson's *The Rise of Professionalism: A Sociological Analysis* (1977). For example, Bledstein argues that contemporary American society is organized by the habits and attitudes appropriate to a "culture of professionalism," which came into existence during the last half of the nineteenth century. The central dynamic of this cultural transformation involved the emergence of a new middle class bent on shaping the world in terms of its own characteristic and unique obsessions with self-discipline, social control, and rational order. The main instrument for reorganizing and reforming society in terms of this new middle class was the modern university that began developing after the Civil War under the guidance of imported German models. Against traditional arguments, which suggest that demands for professional services originate in explosive advances of knowledge and in the conditions of modern society that place a high premium on specialized, technical knowledge, these new sociohistorical analyses emphasize professional self-aggrandizement.

A related body of recent historical studies focuses specifically on the emergence of the helping-healing-human service professions. These include Thomas L. Haskell's book *The Emergence of Professional Social Science* (1977a) and his essay "Power to the Experts" in the *New York Review of Books* (1977b); Christopher Lasch's article on "The Siege of the Family" in the *New York Review of Books* (1977a) and his book entitled *Haven in a Heartless World: The Besieged* (1977b); Anthony M. Platt's study *The Child Savers: The Invention of Delinquency* (1969); and Robert H. Wiebe's *The Search for Order, 1877-1920* (1967). These studies emphasize how the "consensus of the competent"—as Haskell (1977a) refers to the helping-healing-human service professions in his book—came into existence by reducing laymen to dependent incompetence; that is, by transforming self-reliant citizens into (inferior) clients, patients, and consumers of specialized expert services over which laymen have no understanding, authority, or control. This new critique differs from Haug's deprofessionalization argument in several basic respects: it is radical rather than reformist and reality constructionist rather than structural and functionalist. Specifically, the new critique emphasizes the self-aggrandizing monopolization of specialized, technical knowledge by self-interested professional epistemic communities applying social science to family life, the control of deviance, and socialization and resocialization in the post-modern managerial welfare state. These critiques emphasize the professional invention of the very need that professional groups claim to satisfy, as well as the creation or intensification of popular demand for their own services.

These analyses of the social history of the personal and social service professions suggest that the contemporary emphasis on consumerism, consumer participation, and consumer counsels is a ruse on the part of established professions and an intellectual "cop-out" on the part of both liberal reformers and social scientists like Haug (1976). That is, these studies suggest that optimistic accounts of the emergence of intelligent consumers of professional services ignore the power that derives from the monopoly over technical, specialized knowledge on the part of epistemic communities of professional experts—especially those who work in the service agencies of the modern bureaucratic welfare state. This monopoly over specialized, technical knowledge is the basis of their consequent power over their respective client-consumer publics. These studies also represent a new perspective on the professions and professionalization in that they acknowledge the importance as well as the important implications of the fact that the state increasingly pays the bill for most helping-healing-human service professional services—or at least signs the paychecks. In consequence, these analyses devote far more attention than has been frequent in the past to the systematic linkages between contemporary personal service professionalism, social welfare agencies, and vast governmental service bureaucracies.

The general thrust of this recent critical historical literature on the professions, professionalization, and professionalism is to explore the manner in which craft and trade knowledge has been superseded by esoteric, specialized, technical professional knowledge and expertise, as well as the manner in which the autonomous, self-reliant citizen has been superseded by the consumer of professional services who has been transformed into the client-patient—lacking autonomy, self reliance, competence, and common sense—in professional control hierarchies. Thus, these historical studies link the modern professions with broader patterns of control associated with industrial and managerial capitalism, bureaucratic control, and modern production systems.

These recent historically oriented studies of the professions are interesting in several respects. First, they depart from the earlier sociological tradition that typically viewed professionalization and bureaucratization as antithetical tendencies and processes. Instead, these recent historical analyses view both professionalization and bureaucratization as interlocking aspects of a growing centralization in advanced or post-industrial societal structures. Second, this complex of broad social trends is contrasted and counterposed to the forces of populism, localism, community control, structural decentralization, and individual self-help and self-reliance. The characteristic, common to all of these studies, that lends them their radical orientation is their tendency to emphasize and positively value the latter rather than the former characteristics. And what makes them so compatible with the kind of constructionist approach to the sociology of knowledge advanced in this book is their central emphasis on the professional invention of the needs that professional services claim to satisfy, as well as the professional creation and/or intensification of demands for professional services. In an important sense, then, these recent critical historical studies of the professions have revived a scholarly tradition that has been notably absent in the years since Kingsley Davis' article in the first volume of *Psychiatry: The Journal of Interpersonal Studies* (1938) entitled "Mental Hygiene and the Class Structure" and Barbara Wootton's critical masterpiece, *Social Science and Social Pathology* (1959). It is hazardous to predict whether this recent historical interest in critical analyses of the origins of professionalism will expand into a significant body of critical material. However, it does seem clear that the kinds of contemporary issues and problems surrounding the professions and professional autonomy, authority, expertise, knowledge, and status are particularly appropriate subjects, amenable to analysis in terms of the kind of framework and perspective that we have attempted to formulate and develop in this book.

Bibliography

Allardt, Erik, 1971
   "Culture, structure, and revolutionary ideologies." *International Journal of Comparative Sociology* 12:24-40.

Allen, Pamela, 1970
   Free Space. New York: Times Change Press.

Allen, Thomas J. and Stephen I. Cohen, 1969
   "Information flow in research and development laboratories." *Administrative Science Quarterly* 14(March).

American Medical Association, 1953
   It's Your A.M.A. Chicago: Year Book Medical Publishers, Inc.

American Psychiatric Association, 1970
   "Encounter groups and psychiatry." Task Force on Recent Developments in the Use of Small Groups. Washington, D. C. (April).

Anderson, Allen A. and Omar K. Moore, 1969
   "Some principles of the design of clarifying educational environments." Ch. 10 in David A. Goslin (ed.), *Handbook of Socialization Theory and Research.* Chicago: Rand McNally.

Argyris, Chris, 1972
   The Applicability of Organizational Sociology. London and New York: Cambridge University Press.

Aron, Raymond, 1969
   Progress and Disillusion: The Dialectics of Modern Society. New York: Mentor Book, The New American Library.

355

Aubert, Vilhelm and Harrison White, 1959
"Sleep: a sociological interpretation." *Acta Sociologica* IV, 2:46-54 and IV, 3:1-16.

Back, Kurt W., 1970a
"Varieties of sensitivity training." The Sensitivity Training Movement. New York: Russell Sage.

————, 1970b
"Encounter groups and social responsibility." The Sensitivity Training Movement. New York: Russell Sage.

Barber, Bernard, 1963
"Some problems in the sociology of the professions." *Daedalus* XCII (Fall): 672-673.

Bart, Pauline B., 1971
"The myth of a value-free psychotherapy." Pp. 303-327 in Wendell Bell and James A. Mau (eds.), *Sociology and the Future.* New York: Russell Sage.

Batell-Columbus Laboratory, 1973
Interactions of Science and Technology in the Innovative Process: Some Case Studies. Final Report Prepared for the National Science Foundation, Contract NSF C-667 (March 19).

Bauer, Raymond A. (ed.), 1966
Social Indicators. Cambridge, Mass.: The MIT Press.

Baum, Rainer C., 1974
"Beyond convergence." *Sociological Inquiry* 44:225-240.

————, 1975
"The system of solidarities." *Indian Journal of Social Research* XVI, 1-2 (April/August).

Becker, Howard S., 1963
Outsiders, Studies in the Sociology of Deviance. New York: The Free Press of Glencoe.

Bell, Daniel, 1961
"The end of ideology in the west." In Bell (ed.), *The End of Ideology.* New York: Crowell-Collier.

————, 1967a
"Notes on the post-industrial society (I)." *The Public Interest* (Winter).

————, 1967b
"Notes on the post-industrial society (II)." *The Public Interest* (Spring).

Bell, Daniel, 1968
"The measurement of knowledge and technology." Pp. 145–246 in Eleanor B. Sheldon and Wilbert E. Moore (eds.), *Indicators of Social Change, Concepts and Measurements*. New York: Russell Sage.

————, 1973
The Coming of Post-Industrial Society. New York: Basic Books.

————, 1976
The Cultural Contradictions of Capitalism. New York: Basic Books.

————, 1977
"Teletext and technology: new networks of knowledge and information in post-industrial society," *Encounter* XLVIII, 6(June).

Bellah, Robert N., 1964
"Religious evolution." *American Sociological Review* 29(June):358–374.

Ben-David, Joseph, 1960a
"Scientific productivity and academic organization in nineteenth century medicine." *American Sociological Review* 25(December):828–843.

————, 1960b
"Roles and innovation in medicine." *American Journal of Sociology* LXV (May):557–568.

————, 1964
"Professions in the class system of present-day societies." *Current Sociology* 12:247–330.

————, 1968–1969
"The universities and the growth of science in Germany and the United States." *Minerva* 7, 1–2(Autumn-Winter).

————, 1971
The Scientists's Role in Society: A Comparative Study. Englewood Cliffs, N. J.: Prentice-Hall.

Bendix, Reinhard, 1964
"The age of ideology: persistent and changing." Pp. 294–327 in David Apter (ed.), *Ideology and Discontent*. London: Collier-Macmillan.

Bendix, Reinhard and Guenther Roth, 1971
Scholarship and Partisanship: Essays on Max Weber. Berkeley: University of California Press.

Bennis, Warren G., Kenneth D. Benne, and Robert Chin (eds.), 1966
The Planning of Change. New York: Holt, Rinehart and Winston.

Berger, Peter L., 1965
"Psychoanalysis and the sociology of knowledge." *Social Research* (Spring).

————, 1976
"The cultural contradictions of capitalism by Daniel Bell." Book review in *Commentary* 61(April):82–83.

Berger, Peter L. and Thomas Luckmann, 1967
The Social Construction of Reality. Garden City, N. Y.: Doubleday Anchor.

Berger, Peter L., Brigitte Berger, and Hansfried Kellner, 1973
The Homeless Mind: Modernization and Consciousness. New York: Random House.

Bershady, Harold J., 1973
Ideology and Social Knowledge. New York: John Wiley and Sons.

Blain, D., 1953
"Private practice in psychiatry." *The Annals of the American Academy of Political and Social Science* CCLXXXVI(March).

Blau, Peter M., 1974
"Presidential address: parameters of social structure." *American Sociological Review* 39(October):615–635.

Blau, Peter M. and W. Richard Scott, 1962
Formal Organizations: A Comparative Approach. San Francisco: Chandler Publishing Co.

Bledstein, Burton J., 1976
The Culture of Professionalism. New York: Norton.

Blum, Alan F. and Peter McHugh, 1971
"The social ascription of motives." *American Sociological Review* 36(February):98–109.

Blumenfeld, Emily R., 1976
"Child rearing literature as an object of content analysis." *Journal of Applied Communications Research* 4:75–88.

Blumer, Herbert, 1957
"Collective behavior." Pp. 127–158 in J. B. Guttler (ed.), *Review of Sociology: Analysis of a Decade.* New York: John Wiley and Sons.

Bortz, Edward L., 1947
"The objectives of the A.M.A." *Journal of the American Medical Association* 134:565–570.

Bourne, C. P., 1962
"The world's technical journal literature." *American Documentation* 13: 159–168.

Brim, Orville, 1966
"Adult socialization." In Orville Brim and Stanton Wheeler (eds.), *Socialization After Childhood.* New York: Russell Sage.

Brzezinski, Zbigniew, 1968
"America in the technotronic age: new questions of our time." *Encounter* (January):16–23.

Bucher, R., 1962
"Pathology: a study of social movements within a profession." *Social Problems* 10(Summer):45–51.

Caplow, Theodore, 1954
Sociology of Work. Minneapolis: University of Minnesota Press.

Carr-Saunders, A. M., 1966
"Professions: their organization and place in society." Reprinted in Howard M. Vollmer and Donald L. Mills (eds.), *Professionalization*, pp. 3–7. Englewood Cliffs, N. J.: Prentice-Hall.

Cassirer, Ernst, 1944
An Essay on Man: An Introduction to a Philosophy of Human Culture. New Haven: Yale University Press.

Cherniss, Cary, 1972
"Personality and ideology: a personalogical study of women's liberation." *Psychiatry* 35(May):109–125.

Chin, Robert, 1966
"Problems and prospects of applied research." Pp. 667–676 in Bennis, Benne, and Chin (eds.), *The Planning of Change.* New York: Holt, Rinehart and Winston.

Comte, Auguste, 1959
The Positive Philosophy. Tr. H. Martineau. New York: Blanchard.

Crane, Diana, 1969
"Fashion in science: Does it exist?" *Social Problems* 16(Spring):433–441.

————, 1972
Invisible Colleges. Chicago and London: The University of Chicago Press.

Curtis, James E. and John W. Petras (eds.), 1970
The Sociology of Knowledge. New York: Praeger.

De Sola Pool, I., et al. (eds.), 1973
  Handbook of Communication. Chicago: Rand McNally.

Deevey Jr., Edward S., 1960
  "The human population." *Scientific American* (September): 2–9.

Denisoff, R. Serge, 1974
  The Sociology of Dissent. New York: Harcourt Brace Jovanovich.

Dreifus, Claudia, 1973
  Woman's Fate: Raps From a Feminist Consciousness Raising Group. New York: Bantam Books.

Duncan, Hugh D., 1962
  Communication and Social Order. New York and London: Oxford University Press.

Durkheim, Emile, 1914
  "Le dualisme de la nature humaine et ses conditions sociales." Tr. Charles Blend as "The dualism of human nature and social conditions." Reprinted in Kurt H. Wolff (ed.), *Essays on Sociology and Philosophy* by Emile Durkheim et al. Columbus, Ohio: Ohio State University Press, 1960. New York: Harper and Row, Torchbook Edition, 1964.

————, 1915
  The Elementary Forms of the Religious Life. Tr. G. W. Swain. London: Allen and Unwin.

————, 1933
  Division of Labor in Society. Tr. George Simpson. New York: The Free Press of Glencoe.

Durkheim, Emile and Marcel Mauss, 1963
  Primitive Classification. Tr. and ed. Rodney Needham. Chicago: University of Chicago Press.

Eaton, Joseph W. (ed.), 1972
  Institution Building and Development: From Concepts to Applications. Beverly Hills and London: Sage Publications.

Engel, Gloria V., 1970
  "Professional autonomy and bureaucratic organization." *Administrative Science Quarterly* 15(March).

Enzensberger, Hans Magnus, 1974
  The Consciousness Industry. New York: Seabury Press.

Erikson, Erik H., 1963
  Childhood and Society. New York: Norton Books.

Etzioni, A., 1967
"Toward a theory of societal guidance." *American Journal of Sociology* 73(September):173-187.

————, 1968
The Active Society. A Theory of Societal and Political Processes. New York: The Free Press.

Etzioni, A. and Nunn, Richard, 1974
"The public appreciation of science in contemporary America." *Daedalus* 103, 3(Summer).

Ferriss, Abbott L., 1971
"Indicators of trends in the status of American women." New York: Russell Sage Foundation.

Fishbein, Morris, 1947
A History of the A. M. A. Chicago: Year Book Medical Publishers, Inc.

Flacks, Richard, 1970
"Young intelligentsia in revolt." *Transaction* (June):47-55.

Flexner, A., 1910
Medical Education in the United States and Canada. Bulletin 4. New York: Carnegie Foundation for the Advancement of Teaching.

Foucault, Michel, 1972
The Archeology of Knowledge. Tr. A. M. Sheridan Smith. New York: Pantheon Books, Random House.

Foote, Nelson N., 1953
"Professionalization of labor in Detroit." *American Journal of Sociology* LVII(January):372.

Frederick, W. L., 1959
"The history and philosophy of occupational licensing legislation in the United States." *The Journal of the American Dental Association* LVIII, 3(March):18-28.

Freeman, Jo, 1972
"The women's liberation movement: its origins, structures, and ideas." In Hans Peter Freitzel (ed.), *Recent Sociology 4: Family, Marriage, and the Struggle of the Sexes.* New York: The Macmillan Company.

————, 1973
"The origins of the women's liberation movement." *American Journal of Sociology* 78:792-811.

Freidson, Eliot, 1960
"Client control and medical practice." *American Journal of Sociology* LXV(January):382.

————, 1968
"The impurity of professional authority." Ch. 3 in H. S. Becker et al. (eds.), *Institutions and the Person.* Chicago: Aldine.

————, 1970
Profession of Medicine. A Study of the Sociology of Applied Knowledge. New York: Dodd, Mead.

Freidson, Eliot (ed.), 1973
The Professions and Their Prospects. Beverly Hills and London: Sage Publications.

Freud, Sigmund, 1932
The Interpretation of Dreams. London and New York: Hogarth.

Fromm, Erich, 1951
The Forgotten Language: An Introduction to the Understanding of Dreams, Fairy Tales and Myths, New York: Farrar, Straus and Giroux, Inc.

Fuchs, Ralph F., 1964
"Academic freedom—its basic philosophy, function and history." In Hans W. Baade (ed.), *Academic Freedom: The Scholar's Place in Modern Society.* Dobbs Ferry, N. Y.: Oceana Publications, Inc.

Galbraith, John Kenneth, 1967
The New Industrial State. Boston: Houghton Mifflin.

Gallagher, C. F., 1969
"Language rationalization and scientific progress." In Kalman H. Silvert (ed.), *The Social Reality of Scientific Myth.* New York: American Universities Field Staff.

Garceau, Oliver, 1941
The Political Life of the A.M.A. Cambridge Mass.: Harvard University Press.

Garvey, William O., Nan Lin and Carnot E. Nelson, 1970
"Scientific communication in the physical and the social sciences." Baltimore: Johns Hopkins University Center for Research in Scientific Communication. (Mimeographed).

Geertz, Clifford, 1964
"Ideology as a cultural system." Pp. 44-77 in David Apter (ed.), *Ideology and Discontent.* London: Collier-Macmillan.

Geertz, Clifford, 1966

"Religion as a cultural system." Pp. 1–46 in Michael Banton (ed.), *Anthropological Approaches to the Study of Religion.* London: Tavistock Publications, Ltd.

———, 1973

"Common sense as a cultural system." Mimeographed paper prepared for delivery at a conference on Anthropology and Philosophy at Antioch College, May 14–15; subsequently published in the Antioch Review 33, 1(Spring, 1975).

Gehlen, Arnold, 1969

Moral and Hypermoral. Frankfort: Athenaum.

Gieger, Theodore, 1953

Ideologie und Wahrheit, Eine soziologische Kritik des Denkens. Wien: Humboldt Verlag.

Glass, B. and S. H. Norwood, 1959

"How scientists actually learn of work important to them." Proceedings of International Conference on Scientific Information I. Washington: National Academy of Sciences—National Research Council.

Goode, William J., 1957

"Community within a community: The professions." *American Sociological Review* XXII(April):194.

———, 1960

"Encroachment, charlatanism, and the emerging profession: Psychology, sociology and medicine."*American Sociological Review*XXV(December):903.

Goodenough, Ward H., 1957

"Cultural anthropology and linguistics." In P. L. Garvin (ed.), *Report of the Seventh Annual Round Table Meeting on Linguistics and Language Study.* Washington: Georgetown University Monograph Series on Languages and Linguistics, No. 9.

Gornick, Vivian, 1970a

"The new therapies: a brief encounter." *The Village Voice.* (January–February).

———,1970b

"Reflections on collectivities." *Vocations for Social Change* (November–December):28–31.

———, 1971

"Consciousness-raising." *The New York Times Magazine* (January 10).

Gottschalk, L. A. and E. Mansell Pattison, 1969
"Psychiatric perspectives on t-groups and the laboratory movement: an overview." *American Journal of Psychiatry* 126(December):823–839.

Grant, G., 1942
"The guild returns to America." *Journal of Politics* (August):303–336.

Greenwood, Ernest, 1966
"Attributes of a profession." Reprinted in Howard M. Vollmer and Donald L. Mills (eds.), *Professionalization*, pp. 10–17. Englewood Cliffs, N. J.: Prentice-Hall.

Gurvitch, George, 1971
The Social Frameworks of Knowledge. Tr. Margaret A. Thompson and Kenneth A. Thompson. Oxford: Basil-Blackwell and Harper and Row.

——————, 1972
The Social Frameworks of Knowledge. Tr. Margaret A. Thompson and Kenneth A. Thompson. New York: Harper and Row, Torchbook Edition.

Gusfield, Joseph, 1968
"The study of social movements." *The International Encyclopedia of the Social Sciences.* New York: Crowell, Collier and Macmillan.

Hagstrom, Warren O., 1965
The Scientific Community. New York: Basic Books.

Hall, Oswald, 1951
"Sociological research in the field of medicine: progress and prospects." *American Sociological Review* XVI(October):643.

Haskell, Thomas L., 1977a
The Emergence of Professional Social Science. Chicago: University of Illinois Press.

——————, 1977b
"Power to the experts." *New York Review of Books* XXIV, 16(October 13).

Haug, Marie R., 1975
"The deprofessionalization of everyone?" *Sociological Focus* 8(August): 197–213.

Haug, M. R. and M. B. Sussman, 1969
"Professionalism and the public." *Sociological Inquiry* 39(Winter):56–64.

Hayden, Tom, 1970
(Untitled.) *San Francisco Chronicle* (September 15):8.

Heberle, R., 1949
   "Observations on the sociology of social movements." *American Sociological Review* 14:346–357.

Henry, William E., John H. Sims and S. Lee Spray, 1971
   The Fifth Profession: Becoming a Psychotherapist. San Francisco: Jossey-Bass, Inc.

———, 1973
   Public and Private Lives of Psychotherapists. San Francisco: Jossey-Bass, Inc.

Henshel, Richard L., 1975
   "Effects of disciplinary prestige on predictive accuracy: distortions from feedback loops." *Futures* (April):92–106.

Hinkle, Roscoe C., 1976
   "Durkheim's evolutionary conception of social change." *The Sociological Quarterly* 17(Summer):336–346.

Hole, Judith and Ellen Levine, 1970
   Rebirth of Feminism. New York: Quadrangle.

Holton, Gerald (ed.), 1973
   "Modern sciences and the intellectual tradition." *Thematic Origins of Scientific Thought.* Cambridge, Mass.: Harvard University Press.

Holzner, Burkart, 1965
   "Observer and agent in the social process." *Archives of Social and Legal Philosophy* LII.

———, 1968
   Reality Construction in Society. Cambridge, Mass.: Schenkman Publishing Company.

———, 1972
   Reality Construction in Society. Revised Edition. Cambridge, Mass.: Schenkman Publishing Company.

———, 1973a
   "Sociology and the consciousness of society." *Sociologia Internationales* 11.

———, 1973b
   "The uses of sociological theory." Paper presented at Greystone Conference on Knowledge Utilization. New York.

Holzner, Burkart, Evelyn Fisher and John Marx, 1977
"Paul Lazarsfeld and the study of knowledge applications." *Sociological Focus 10,* 2(April).

Holzner, Burkart and Leslie Salmon-Cox, 1977
"Conceptions of research and development for education in the United States." The Annals of the Academy of Social and Political Science (November).

Homans, George C., 1950
The Human Group. New York: Harcourt Brace.

Hughes, Everett C., 1958
Men and their Work. New York: The Free Press of Glencoe.

Hughes, H. Stuart, 1958
Consciousness and Society. Glencoe, Ill.: The Free Press.

Human Interaction Research Institute, 1976
Putting Knowledge to Use: A Distillation of the Literature Regarding Knowledge Transfer and Change. Los Angeles: Human Interaction Research Institute, NIMH, Rockville, Md.

Hyde, David R., W. Wolff, and P. Payson, 1954
"The American medical association: Power, purpose, and politics of organized medicine." *Yale Law Journal* 63:938–1021.

Imersheim, Allen W., 1977
"The epistemological bases of social order: toward ethno-paradigm analysis." Pp. 1–51 in David R. Heise (ed.), *Sociological Methodology.* San Francisco: Jossey-Bass, Inc.

Jacobs, Ruth, 1971
"Emotive and control groups as mutated new American utopian communities." *Journal of Applied Behavioral Science* 7:2.

James, William, 1952
The Principles of Psychology. Chicago: Encyclopedia Britannica Co.

Jellinek, E. M., 1960
The Disease Concept of Alcoholism. New Haven: Hillhouse Press.

Jonas, Hans, 1959
"The practical uses of theory." *Social Research* 26(Summer):127–166.

————, 1963
"The practical uses of theory." In Maurice A. Natanson (ed.), *Philosophy of the Social Sciences.* New York: Random House.

Kagan, J., 1969
  "Continuity in development." Paper presented at the meeting of the Society for Research in Child Development. Santa Monica, California.

Kalman, H. Silvert, 1969
  The Social Reality of Scientific Myth: Science and Social Change. New
  York: American Universities Field Staff.

Kant, Immanuel, 1964
  Werke VI. Insel Verlag.

Kanter, Rosabeth Moss, 1968
  "Commitment in social organization: a study of commitment mechanisms
  in utopian communities." *American Sociological Review* 33(August):499-
  517.

Katz, F. E., 1968
  Autonomy and Organization. The Limits of Social Conflict. New York:
  Random House.

Kavolis, Vytautas, 1968
  Artistic Expression: A Sociological Analysis. Ithaca, New York: Cornell
  University Press.

————, 1969
  "Revolutionary metaphors and ambiguous personalities: notes toward an
  understanding of post-modern revolutions." *Soundings* 52(Winter):394-
  414.

————, 1970
  "Post-modern man: psycho-cultural responses to social trends." *Social
  Problems* 17(Spring):435-449.

————, 1974
  "Notes on post-industrial culture." *Arts in Society* 11(Fall-Winter).

————, 1975
  "Logics of selfhood and modes of order: civilizational structures for
  individual identities." A paper presented at the Conference on Identity
  and Authority. University of Pittsburgh (April).

Keniston, Kenneth, 1968-1969
  "Heads and seekers: drugs on campus, counter-cultures and American
  society." *The American Scholar* 38(Winter):97-112.

King, Richard, 1972
  "The eros ethos cult in the counter-culture." *Psychology Today* (August):
  35-70.

Kleitman, Nathaniel, 1939
    Sleep and Wakefulness. Chicago: University of Chicago Press.

Kornhauser, William, 1962
    Scientists in Industry: Conflict and Accommodation. Berkeley and Los
    Angeles: University of California Press.

Kroeber, Alfred and Talcott Parsons, 1958
    "The concept of culture and of social system." *American Sociological Re-
    view* 23:582-583.

Krohn, Roger G., 1972
    "Patterns of the institutionalization of Research." In Saad Z. Nagi and
    Ronald G. Corwin (eds.), *The Social Contexts of Research.* New York and
    Toronto: Wiley Interscience, Division of John Wiley and Sons, Inc.

Kuhn, Thomas S., 1962
    The Structure of Scientific Revolutions. Chicago: University of Chicago
    Press.

————, 1970
    The Structure of Scientific Revolutions. Revised Edition. Chicago: Uni-
    versity of Chicago Press.

————, 1974
    "Second thoughts on paradigms." In Frederick Suppes (ed.), *The Struc-
    ture of Scientific Theories.* Chicago: University of Chicago Press.

Lakin, Martin, 1969
    "Some ethical issues in sensitivity training." *American Psychologist* 24
    (October):923-928.

————, 1971
    "Group sensitivity training and encounter—time out for measurement."
    Unpublished manuscript.

Lane, R. E., 1966
    "The decline of politics and ideology in a knowledgeable society." *Ameri-
    can Sociological Review* 31(October):649-662.

Langer, Susan, 1951
    *Philosophy in a New Key: A Study in the Symbolism of Reason, Rite and
    Art.* New York: The New American Library, Mentor Books.

Lasch, Christopher, 1977a
    "The siege of the family." *New York Review of Books* XXIV, 19(Novem-
    ber 24).

——————, 1977b
Haven in a Heartless World: The Family Beseiged. New York: Basic Books.

Lazarsfeld, Paul and Jeffrey G. Reitz in collaboration with Ann K. Pasanella, 1975
An Introduction to Applied Sociology. New York: Elsevier.

Levinson, D. J., 1967
"Medical education and the theory of adult socialization." *Journal of Health and Social Behavior.*

Lévi-Strauss, Claude, 1963
Structural Anthropology. Tr. Claire Jacobson and Brooke Grundfest Schoepf. New York: Basic Books.

Lewin, B. and H. Ross, 1960
Psychoanalytic Education in the United States. New York: Norton.

Lindsmith, A. R. and A. L. Strauss, 1957
Social Psychology. New York: The Dryden Press.

Löwith, Karl, 1964
From Hegel to Nietsche: The Revolution in Nineteenth Century Thought. Tr. David E. Green. New York: Holt, Rinehart and Winston.

Lundberg, Craig C., 1966
"Middle-men in science utilization: some notes toward clarifying conversion roles." *American Behavioral Scientist* 9(February):11–14.

MacDermot, H. E., 1935
The History of the Canadian Medical Association. Toronto: University of Toronto Press.

Machlup, Fritz, 1962
The Production and Distribution of Knowledge in the United States. Princeton, N. J.: Princeton University Press.

MacIver, Robert M., 1955
Academic Freedom in Our Time. New York: Norton.

——————, 1966
"Social significance of professional ethics." Reprinted in Howard M. Vollmer and Donald L. Mills (eds.), *Professionalization,* pp. 51–54. Englewood Cliffs, N. J.: Prentice-Hall.

Magraw, R. M., 1966
Ferment in Medicine: A Study of the Essence of Medical Practice and of Its New Dilemmas. Philadelphia: W. B. Saunders Co.

McGraw, R. M., 1971
"Trends in medical education and health services: their implications for a career in family medicine." *New England Journal of Medicine* 285:1407–1413.

Maisel, Richard, 1972
Information Technology. New York: Conference Board Inc.

Maliver, Bruce L., 1971
"Encounter groups up against the wall." *The New York Times Magazine* (January 3).

Mannheim, Karl, 1936
Ideology and Utopia: An Introduction to the Sociology of Knowledge. Tr. Louis Wirth and Edward A. Shils. New York: Harcourt, Brace and World.

————, 1940
Man and Society in an Age of Reconstruction: Studies in Modern Social Structure. Tr. Edward Shils. London: Routledge and Keegan Paul. New York: Harcourt, Brace and World.

————, 1943
Diagnosis of our Time: War Time Essays of a Sociologist. London: Keegan Paul, Trench, Trubner. New York: Oxford University Press, 1944.

————, 1950
Freedom, Power and Democratic Planning. Ernest K. Bramsted and Hans Gerth (eds.). New York: Oxford University Press. London: Routledge and Keegan Paul, 1951

————, 1952
Essays in the Sociology of Knowledge. Paul Kecskemeti (ed.). London: Routledge and Keegan Paul.

Manuel, Frank E., 1965
"Toward a psychological history of utopias." *Daedalus* XCIV: 293–322.

————, 1976
"The cultural contradictions of capitalism by Daniel Bell." Book review in *The New Republic* (March 20).

Marin, Peter, 1975
"The new narcissism." *Harper's* (October):45–56.

Marquis, Donald G. and Thomas J. Allen, 1966
"Communication patterns in applied technology." *American Psychologist* 21(November).

Marx, John H., 1969
"A multidimensional conception of ideologies in professional arenas: the case of the mental health field," *Pacific Sociological Review* 12(Fall):75-85.

—————, 1979
"The ideological construction of post-modern identities in contemporary cultural movements: pursuing an 'infinite possibility thing.'" In Roland Robertson and Burkart Holzner (eds.), *Identity and Authority*. London: Basil-Blackwell, 1979.

Marx, John H. and David Ellison, 1975
"Sensitivity training and communes: contemporary quests for community." *Pacific Sociological Review*.

Marx, John H. and Burkart Holzner, 1975
"Ideological primary groups in contemporary cultural movements." *Sociological Focus* 8(October):311-329.

—————, 1977
The social construction of strain and ideological models of grievance in contemporary movements." *Pacific Sociological Review* 20, 3(July).

Marx, John H., Patricia Rieker, and David L. Ellison, 1974
"The sociology of community mental health: historical and methodological perspectives." Pp. 9-41 in Paul M. Roman and Harrison M. Trice (eds.), *Sociological Perspectives on Community Mental Health*. Philadelphia: F.A. Davis Company.

Marx, John H. and Joseph H. Seldin, 1973a
"At the crossroads of crisis I: therapeutic sources and quasi-therapeutic functions of post-industrial communes." *Journal of Health and Social Behavior* 14(March):39-52.

—————, 1973b
"At the crossroads of crisis II: organizational and ideological bases of contemporary communes." *Journal of Health and Social Behavior* 14(June): 183-191.

Marx, John H. and S. Lee Spray, 1970
"Marital status and occupational success among mental health professionals." *Journal of Marriage and the Family* 32 (February):110-118.

—————, 1972
"Psychotherapeutic 'birds of a feather': social class status and religio-cultural value homophily in the mental health field." *Journal of Health and Social Behavior* 13(December):413-428.

Marx, Karl, 1904

> *A Contribution to the Critique of Political Economy.* Tr. N. I. Stone. 2nd German Edition, Chicago: Kerr.

————, 1964

> "Capital, Vol. 1 (1867)." Quoted from Karl Marx in Bottomore and Ruble (eds.), *Selected Writings in Sociology and Social Philosophy.* Tr. T. B. Bottomore. N.J.: McGraw-Hill Book Company.

————, 1968

> *Theses on Feuerbach.* See, e.g., "Inter nationes Karl Marx 1818/1968." Cologne: Rudolf Miller, p. 145ff.

Marx, Karl and Friedrich Engels, 1939

> *The German Ideology.* Ed. R. Pascoe. London: Lawrence and Wishart.

Mauss, Armand L. and Donald W. Peterson, 1973

> "The cross and the commune: an interpretation of the Jesus people." In Charles Y. Glock (ed.), *Religion in Sociological Perspective.* Belmont, California: Wadsworth Publishing Co.

McLuhan, Marshall, 1962

> *The Gutenberg Galaxy.* Toronto, Ontario: University of Toronto Press.

————, 1967

> *The Medium is the Message.* New York: Bantam Books.

McQuail, Denis (ed.), 1972

> *Sociology of Mass Communications.* London: Penguin.

Menzel, Herbert, 1962

> "Planned and unplanned scientific communication." In Bernard Barber and Walter H. Hirsch (eds.), *The Sociology of Science.* New York: The Free Press.

————, 1966

> "Scientific communication: Five themes from social science research." *American Psychologist* 21(November).

————, 1976

> "Quasi-communication." *Public Opinion Quarterly* 5, 3.

Merton, Robert K., 1938

> "Science, technology and society in seventeenth century England." *Ouris* IV:360-632.

————, 1957, 1968

> *Social Theory and Social Structure.* New York: The Free Press.

Merton, Robert K., 1963
"The ambivalence of scientists." *Bulletin* 112 (February): 77–97.

————, 1969
"Behavior patterns of scientists." *American Scientist* 57(Spring):1–23.

————, 1973
The Sociology of Science: Theoretical and Empirical Investigations. Ed. Norman W. Storer. Chicago: The University of Chicago Press.

———— and Eleanor Barber, 1976, 1963
"Sociological ambivalence." Pp. 91–120 in E. A. Tiryakian (ed.). *Sociological Theory, Values and Sociocultural Change.* Glencoe, Ill.: The Free Press.

Miller, George A., 1956
"The magical number seven, plus or minus two—some limits on our capacity for processing information." *Psychological Review* 63.

Mills, C. Wright, 1951
White Collar: The American Middle Classes. New York, N. Y.: Oxford University Press.

Mischel, Walter, 1969
"Continuity and change in personality." *American Psychologist* 24:1013–1018.

Mitchell, Juliet, 1971
Woman's Estate. New York: Pantheon Books, Random House.

Mitroff, Ian, 1972
The Subjective Side of Science: A Philosophical Inquiry into the Psychology of the Apollo Moon Scientists. Amsterdam and New York: Elsevier.

————, 1974
"Norms and counter-norms in a select group of apollo moon scientists: A case study of the ambivalence of scientists." *American Sociological Review* 38(June):579–595.

Moore, Omar K., 1973
"The science of knowledge." In The Sociological Implications of Sociological Change, a four-part Cogar Foundation Series, presented at Herkimer County Community College, first lecture (November).

Moore, W. E., 1970
The Professions: Rules and Roles. New York: Russell Sage.

Ms. Magazine
A Guide to Consciousness Raising. New York: *Ms.* Magazine.

Mueller, Claus, 1973
   The Politics of Communication. New York and London: Oxford University Press.

Mullins, Nicholas C., 1968
   "The distribution of social and cultural properties in informal communication networks among biological scientists." *American Sociological Review* 33, 5(October):786–797.

————, 1973
   Theories and Theory Groups in Contemporary American Sociology. New York: Harper and Row.

Nadel, Siegfried Frederick, 1957
   The Theory of Social Structure. London: Cohen and West.

Nahirny, Vladimir C., 1962
   "Some observations on ideological groups." *The American Journal of Sociology* 68:173–181.

Nehnevajsa, Jiri, 1972
   "Methodological issues in institution building." In Joseph Eaton (ed.), *Institution Building and Development.* Beverly Hills and London: Sage Publications.

Nelson, Richard, Merton J. Peck and Edward D. Kalachek, 1967
   Technology, Economic Growth and Public Policy. A Rand Corporation and Brookings Institution Study. Washington: Brookings Institution.

Nisbet, R. A., 1952
   "Conservatism and sociology." *American Journal of Sociology* 58:167–175.

Oberschall, Anthony, 1973
   Social Conflict and Social Movements. Englewood Cliffs, N. J.: Prentice-Hall.

Oppenheimer, Robert, 1963
   "Communication and comprehension of scientific knowledge." *Science* 142(November).

Ornstein, Martha, 1928
   The Role of Scientific Societies in the Seventeenth Century. Chicago: University of Chicago Press.

Parsons, Talcott, 1949
   Essays in Sociological Theory: Pure and Applied. New York: The Free Press.

————, 1950
   The Social System. Glencoe, Ill.: The Free Press.

Parsons, Talcott, 1954
"A sociologist looks at the legal professions." In Talcott Parsons, *Essays in Sociological Theory: Pure and Applied.* New York: The Free Press.

————, 1964
"Definitions of health and illness in the light of American values and social structure." In Talcott Parsons (ed.), *Social Structure and Personality.* New York: The Free Press.

Parsons, Talcott and Renee Fox, 1952
"Illness and the urban American family." *Journal of Social Issues* 8:31–44.

Parsons, Talcott, Renee C. Fox and Victor M. Lidz, 1972
"The gift of life and its reciprocation." *Social Research* 39:367–415.

Pelz, Donald C. and Frank M. Andrews, 1966
Scientists in Organizations. New York: John Wiley & Co.

Platt, Anthony M., 1969
The Child Savers: The Invention of Delinquency. Chicago: University of Chicago Press.

Price, Derek J. De Solla, 1961
Science Since Babylon. New Haven: Yale University Press.

————, 1963
Little Science, Big Science. New York and London: Columbia University Press.

————, 1965
"Is technology historically independent of science?" *Technology and Culture* 6.

Price, Derek J. De Solla and Donald D. B. Beaver, 1966
"Collaboration in an invisible college." *American Psychologist* 21(November).

Radnor, Michael, Earl C. Young and Harriet Spivak, 1975
"Analysis of comparative research, development and innovation systems and management: With implications for education." Research Report for the National Institute of Education (April).

Radnor, Michael, Harriet Spivak, Durward Hofler, in collaboration with Earl C. Young, 1976
"Agency-field relationships in the educational R/D&I systems: a policy analysis for the national institute of education." Evanston, Ill.: Northwestern University, Center for the Interdisciplinary Study of Science and Technology (October).

Ransom, Donald C. and Herbert E. Vandervoort, 1973
"The development of family medicine, problematic trends." *Journal of the American Medical Association.* 225, 9(August).

Rayack, E., 1967
Professional Power and American Medicine. The Economics of the American Medical Association. Cleveland: World.

Reich, Charles, 1970
The Greening of America. New York: Random House.

Reiff, Philip, 1966
The Triumph of the Therapeutic. New York: Harper Torchbooks.

Remmling, Gunter W., 1967
Road to Suspicion: A Study of Modern Mentality and the Sociology of Knowledge. New York: Appleton-Century-Crofts.

————, 1973
Towards the Sociology of Knowledge: Origin and Development of a Sociological Thought Style. New York: Humanities Press.

Richmond, J. B., 1969
Currents in American Medicine: A Developmental View of Medical Care and Education. Commonwealth Fund Book. Cambridge: Harvard University Press.

Ritzer, George, 1975
Sociology, A Multiple Paradigm Science. Boston: Allyn & Bacon.

Robertson, Roland and Burkart Holzner (eds.), forthcoming
Identity and Authority. London: Basil-Blackwell, 1979.

Rogers, E. M., 1962
Diffusion of Innovations. New York: Free Press of Glencoe.

Rogers, Everett M. and F. Floyd Shoemaker, 1971
Communication of Innovations. 2nd Edition. New York: The Free Press.

Rosenberg, Dorothy B., 1977
"Acupuncture and U.S. medicine: A social historical study of the response to the availability of knowledge." Ph.D. dissertation. University of Pittsburgh.

Roszak, Theodore, 1969
The Making of a Counter-Culture. New York: Random House.

Rowe, A. P., 1964
"From scientific idea to practical use." *Minerva* 2(Spring).

Ruitenbeek, Hendrik M., 1970
The New Group Therapies. New York: Discus Books.

Scheff, Thomas J., 1966
Being Mentally Ill: A Sociological Theory. Chicago: Aldine Publishing Company.

Scheler, Max, 1954
The Nature of Sympathy. Tr. Peter Heath. New Haven: Yale University Press.

————, 1960
Die Wissensformen und die Gesellschaft. Bern: Francke.

Schramm, W. (ed.), 1963
The Science of Human Communication. New York: Basic Books.

Schumpeter, Joseph, 1929
Das Soziale Antlitz des Deutschen Reiches. University of Bonn: Bonner Mitteilungen.

Schutz, Alfred, 1962
Collected Papers, Vol. I. *The Problem of Social Reality.* Ed. Maurice A. Natanson. The Hague: Martinues Nijhoff.

————, 1967
The Phenomenology of the Social World. Tr. George Walsh and Frederick Lenhert. Evanston, Ill.: Washington University Press.

Sherif, M., 1952
"The concept of reference groups in human relations." Pp. 203-231 in M. Sherif and M.O. Wilson (eds.), *Group Relations at the Crossroads.* New York: Harper and Brothers.

Sigerist, H. E., 1935
"A history of medical licensure." *The Journal of the American Medical Association* CIV(March):1057-1060.

Simmel, Georg, 1950
The Sociology of Georg Simmel. Tr. Kurt H. Wolff. New York: The Free Press.

Sisk, John P., 1976
"Salvation unlimited." *Commentary* 61(April):52-56.

Slater, Philip E., 1970
The Pursuit of Loneliness: American Culture at the Breaking Point. Boston: Beacon Press.

Sloan, Alfred P., Jr., 1965
   My Years with General Motors. Ed. John McDonald with Catherine Stevens. Garden City: Doubleday.

Smelser, Neil J., 1963
   Theory of Collective Behavior. New York: The Free Press.

Spence, Michael A., 1974
   "An economist's view of information." In Carlos A. Guadra and Ann W. Luke (eds.), *Annual Review of Information Science and Technology 9.* Washington, D.C.: American Society for Information Science.

Spradley, James P., 1972a
   Culture and Cognition: Rules, Maps, and Plans. San Francisco and Toronto: Chandler Publishing Company.

————, 1972b
   "Foundations of cultural knowledge." Pp. 3–38 in James P. Spradley (ed.), *Culture and Cognition.* San Francisco and Toronto: Chandler Publishing Company.

Steensgaard, Niels, 1976
   Informal paper presented at the annual meeting of the International Society for the Comparative Study of Civilizations. Philadelphia, Pa.: University of Pennsylvania (April 1–4).

Stone, Anthony R., 1969
   "The interdisciplinary research team." *The Journal of Applied Behavioral Science* 5(July–August–September).

Strauss, A., 1947
   "Research in collective behavior: Neglect and need." *American Sociological Review* 12:352–354.

Swanson, Guy E., 1979
   "A basis of authority and identity in post-industrial society." In Roland Robertson and Burkart Holzner (eds.), *Identity and Authority.* London: Basil-Blackwell. 1979.

Szasz, Thomas S., 1964
   The Myth of Mental Illness. New York: Harper and Row.

Tanner, Leslie B., 1970
   Voices from the Women's Liberation Movement. New York: New American Library.

Thomas, W. I., 1923
   The Unadjusted Girl. New York: Harper Torchbooks, 1967.

Timasheff, N. S., 1966
"Business and the professions in liberal, fascist and communist society." Reprinted in Howard M. Vollmer and Donald L. Mills (eds.), *Professionalization*, p. 59. Englewood Cliffs, N. J.: Prentice-Hall.

Tiryakian, Edward A., 1975
"Neither Marx nor Durkheim . . . perhaps Weber." *American Journal of Sociology* 81, 1(July):1–33.

Toffler, Alvin W., 1970
Future Shock. New York: Random House.

Touraine, Alain, 1972
The Post-Industrial Society, Tomorrow's Social History: Classes, Conflicts, and Culture in the Programmed Society. Tr. Leonard F. X. Mayhew. New York: Random House.

Turner, Ralph H., 1969
"The theme of contemporary social movements." *British Journal of Sociology* 20(December):586–599.

U. S. Department of Commerce, Bureau of the Census, 1975
Historical Statistics of the United States, Colonial Times to 1970. Bicentennial Edition, Parts 1 and 2. 93rd Congress, 1st Session, House Document no. 93–78.

Vollmer, Howard M. and Donald L. Mills (eds.), 1966
Professionalization. Englewood Cliffs, N. J.: Prentice-Hall.

Wallace, Anthony F. C., 1956
"Revitalization movements." *American Anthropologist* 58:264–281.

————, 1961a
Culture and Personality. New York: Random House.

————, 1961b
"On being just complicated enough." Proceedings of National Academy of Science 47:458–464.

————, 1962
"Culture and cognition." *Science* 135:351–357.

————, 1972
"Driving to work." Pp. 310–326 in James P. Spradley (ed.), *Culture and Cognition*. San Francisco and Toronto: Chandler Publishing Company.

Wallace, Anthony F. C. and John Atkins, 1960
"The meaning of kinship terms." *American Anthropologist* 62:58–80.

Weber, Max, 1949, 1956
  The Methodology of the Social Sciences. Tr. and ed. Edward A. Shils and
  Henry A. Finch. Glencoe, Ill.: The Free Press.

————, 1956
  Wirtschaft und Gesellschaft. 2 vols. Ed. Johannes Winckelmann Köln
  Kiepenhener und Witsch.

————, 1958a
  "Science as a vocation." P. 129 in Hans W. Gerth and C. Wright Mills
  (eds.), *From Max Weber.* New York: Oxford University Press.

————, 1958b
  The Protestant Ethic and the Spirit of Capitalism. Tr. Talcott Parsons with
  a Foreword by R. H. Tawney. New York: Charles Scribners Sons.

————, 1963
  Gesammelte Aufsätze zur Religions-Soziologie. 5th Ed. Vol. I–III. Tü-
  bingen, J. C. B. Mohr (Paul Siebeck).

————, 1968
  Economy and Society: An Outline of Interpretive Sociology. 3 vols.
  Ed. Guenter Roth and Claus Wittich. Tr. Ephraim Fischoff et al. New
  York: Bedminster Press.

Weinberg, A. M., 1967
  "Social problems and national socio-technical institutes." *Applied Science
  and Technological Progress.* Report to the Committee on Science and
  Astronautics, U. S. House of Representatives, National Academy of
  Sciences Report. Washington, D. C. (June).

Weller, Jack M. and E. L. Quarantelli, 1973
  "Neglected characteristics of collective behavior." *American Journal of
  Sociology* 79:665–685.

Wheeler, Stanton, 1966
  "Socialization in institutions." In Orville Brim and Stanton Wheeler (eds.),
  Socialization After Childhood. New York: Russell Sage.

Wheelis, Allen, 1953
  The Quest for Identity. New York: Norton.

Wiebe, Robert H., 1967
  The Search for Order, 1877–1920. San Francisco: Hill and Wang.

Wilensky, Harold, 1964
  "The professionalization of everyone?" *American Journal of Sociology*
  70 (September):137–158.

Wilensky, Harold L. and Jack Ladinsky, 1967
"From religious community to occupational group: structural assimilation among professors, lawyers, and engineers." *American Sociological Review* 32(August):541–561.

Willard, W., 1966
Meeting of the Challenge of Family Practice. Chicago: American Medical Association.

Woods, Ralph L., 1947
The World of Dreams: An Anthology. New York: Farrar, Straus and Giroux, Inc.

Wootton, Barbara, 1959
Social Science and Social Pathology. London: George Allen and Unwin.

Zinman, John, 1968
Public Knowledge: The Social Dimension of Science. New York: Cambridge University Press.

Index